Peter T. Johnstone
April 2000.

Power Algebras over Semirings

Mathematics and Its Applications

Managing Editor:

M. HAZEWINKEL

Centre for Mathematics and Computer Science, Amsterdam, The Netherlands

Volume 488

Power Algebras over Semirings

With Applications in Mathematics and Computer Science

by

Jonathan S. Golan
Department of Mathematics,
University of Haifa,
Haifa, Israel

KLUWER ACADEMIC PUBLISHERS
DORDRECHT / BOSTON / LONDON

Library of Congress Cataloging-in-Publication Data

ISBN 0-7923-5834-1

Published by Kluwer Academic Publishers,
P.O. Box 17, 3300 AA Dordrecht, The Netherlands.

Sold and distributed in North, Central and South America
by Kluwer Academic Publishers,
101 Philip Drive, Norwell, MA 02061, U.S.A.

In all other countries, sold and distributed
by Kluwer Academic Publishers,
P.O. Box 322, 3300 AH Dordrecht, The Netherlands.

Printed on acid-free paper

All Rights Reserved
© 1999 Kluwer Academic Publishers
No part of the material protected by this copyright notice may be reproduced or
utilized in any form or by any means, electronic or mechanical,
including photocopying, recording or by any information storage and
retrieval system, without written permission from the copyright owner.

Printed in the Netherlands.

to Theresa,

whose friendship was everything

Table of Contents

Preface	ix
Some (hopefully) motivating examples	1
Chapter 0: Background material	7
Chapter 1: Powers of a semiring	27
Chapter 2: Relations with values in a semiring	37
Chapter 3: Change of base semirings	61
Chapter 4: Convolutions	67
Chapter 5: Semiring-valued subsemigroups and submonoids	89
Chapter 6: Semiring-valued groups	113
Chapter 7: Semiring-valued submodules and subspaces	141
Chapter 8: Semiring-valued ideals in semirings and rings	155
References	169
Index	189

Preface

This monograph is a continuation of several themes presented in my previous books [146, 149]. In those volumes, I was concerned primarily with the properties of semirings. Here, the objects of investigation are sets of the form R^A, where R is a semiring and A is a set having a certain structure. The problem is one of translating that structure to R^A in some "natural" way. As such, it tries to find a unified way of dealing with diverse topics in mathematics and theoretical computer science as formal language theory, the theory of fuzzy algebraic structures, models of optimal control, and many others. Another special case is the creation of "idempotent analysis" and similar work in optimization theory. Unlike the case of the previous work, which rested on a fairly established mathematical foundation, the approach here is much more tentative and docimastic. This is an introduction to, not a definitive presentation of, an area of mathematics still very much in the making.

The basic philosphical problem lurking in the background is one stated succinctly by Höhle and Šostak [185]:

"... to what extent basic fields of mathematics like algebra and topology are dependent on the underlying set theory?"

The conflicting definitions proposed by various researchers in search of a resolution to this conundrum show just how difficult this problem is to see in a proper light. I try to acknowledge this fact by often indicating the existence routes which I do not intend to pursue at this moment, but which should serve as an invitation and challenge to the reader.

Since the development of a new mathematical theory is basically inductive – one begins with a large number of instances which appear in various mathematical contexts and tries to develop a general abstract framework in which to understand them best, I have tried to give a large number of examples, taken from various areas, to illustrate where the ideas here originated. These are not pursued in detail, but citations to the literature will allow those interested to study them in greater depth.

Much of the material in this work was originally arranged for presentation at a seminar on semirings which I directed while a visiting professor at the University of Idaho during the 1997/8 academic year. In addition to me, the participants in the seminar included professors Erol Barbut and Willy Brandal, and the graduate students Lixin Huang, and Minglong Wu. I owe them – and the various other faculty members and students whom, from time to time, I managed to waylay in the halls – many thanks for their comments and suggestions and for patiently following me down several mathematical dark alleys in the search of the tao of power algebras. The references chosen were intended to illustrate my approach, and should in no way be considered a comprehensive survey of the literature. In all probability, more relevant results have been left out than have been included.

Special thanks are due to the University of Haifa for granting me sabbatical leave during the 1997/8 academic year and to the Department of Mathematics at the University of Idaho for their warm hospitality and for arranging, and partially funding, my very enjoyable stay in that beautiful and tranquil area of the United States. Similarly, thanks are due to the Departments of Mathematics at the University of Minnesota in Minneapolis/St. Paul, the University of Tennessee in Knoxville, and Rutgers University in Piscataway, for arranging, and partially funding, my stay at their respective institutions during the summer of 1998.

Some (Hopefully) Motivating Examples

In order to motivate the topics discussed in this volume, we begin with several examples.

Let A be a nonempty set. It is well-known that there is a bijective correspondence between the family of all subsets of A and the family $\mathbb{B}^A$ of all functions from A to $\mathbb{B} = \{0, 1\}$, which assigns to each subset B of A its *characteristic function*

$$\chi_B : a \mapsto \begin{cases} 1 & \text{if } a \in B \\ 0 & \text{otherwise.} \end{cases}$$

The set $\mathbb{B}$ of course, has the structure of a complete bounded distributive (in fact linear) lattice with respect to the operations $\vee$ (supremum) and $\wedge$ (infimum) and with the induced partial order $\leq$ given by $0 \leq 1$. If we think of $\mathbb{B}^A$ as a direct product of copies of $\mathbb{B}$, then we see that these operations and this partial order carry over to $\mathbb{B}^A$ by componentwise definition:

(1) If $U \subseteq \mathbb{B}^A$ and if $a \in A$ then

$$(\vee U) : a \mapsto \vee \{\chi(a) \mid f \in U\}$$

and

$$(\wedge U) : a \mapsto \wedge \{\chi(a) \mid f \in U\}$$

(2) If $f, f' \in \mathbb{B}^A$ then $f \leq f'$ if and only if $f(a) \leq f'(a)$ for all $a \in A$.

As a consequence of these definitions, it is easy to see that if $\{B_i \mid i \in \Omega\}$ is a family of subsets of A and if we set $C = \cup_{i \in \Omega} B_i$ and $D = \cap_{i \in \Omega} B_i$ then

$$\bigvee_{i \in \Omega} \chi_{B_i} = \chi_C$$

and
$$\bigwedge_{i \in \Omega} \chi_{B_i} = \chi_D.$$

Moreover, if B and B' are subsets of A then $B \subseteq B'$ if and only if $\chi_B \leq \chi_{B'}$.

These observations, which should be familiar to every undergraduate mathematics major, lead to interesting generalizations, which have their origins in several variations on simple Cantorian set theory. These took several forms:

(I) EXAMPLE. It is sometimes very important to allow an element of a set to appear in that set "more than once". This has led to the theory of *multisets*, which were first introduced by Donald Knuth [209] for use in computer science and have since been used extensively in many contexts. Thus, given a nonempty set A, a multisubset of A is defined by a *multiplicity function* in $\mathbb{N}^A$, where $\mathbb{N}$ is the set of all nonnegative integers. The theory of multisets has been formalized in [42]. For a formalization of linear logic in terms of multisets, refer to [13] and [382].

(II) EXAMPLE. Loeb [251], concerned with various combinatorial problems, extended the notion of a multiset to that of a *hybrid set*, or "set with a negative number of elements" by considering multiplicity functions belonging to $\mathbb{Z}^A$. Also refer to [71]. For the use of hybrid sets in the construction of colored Petri nets, refer to [191]. Another extension of the notion of a multiset involves looking at multiplicity functions in R^A, where $R = \mathbb{N} \cup \{-\infty, \infty\}$ and where the usual addition and multiplication of $\mathbb{N}$ are augmented in the following manner:

$$-\infty + r = r + (-\infty) = -\infty \text{ for all } r \in R;$$
$$\infty + r = r + \infty = \infty \text{ for all } -\infty \neq r \in R;$$
$$r \cdot 0 = 0 \cdot r = 0 \text{ for all } r \in R;$$
$$-\infty \cdot r = r \cdot -\infty = -\infty \text{ for all } 0 \neq r \in R;$$
$$\infty \cdot r = r \cdot \infty = \infty \text{ for all } 0, -\infty \neq r \in R.$$

Elements of R^A are sometimes called *bags* on A. (On the other hand, the term "bag" is often used as a synonym for "multiset", so one has to be careful.) See [19] for an application of this construction to signal processing and [411] for an application to the modeling of fuzzy systems.

(III) EXAMPLE. Zadeh [416], in his ground-breaking work, enlarged the concept of a subset further. Given a nonempty set A, a *fuzzy subset* of A is defined by an *extent of membership function* in $\mathbb{I}^A$, where $\mathbb{I}$ is the unit interval on the real line. The theory of fuzzy sets, which has since spawned an extremely large mathematical and engineering literature and which has led to many interesting and significant real-world applications, allowed a mathematical treatment of situations in which the extent of subset membership may be known only roughly or approximately. It is important to distinguish fuzziness from probablity: basically, fuzziness describes the ambiguity or uncertainty of events; probability describes the occurrence of events.

In the original conception, the unit interval $\mathbb{I}$ was equipped with the operations $\vee$ (maximum) and $\wedge$ (minimum). An important extension of this situation, however, was realized when the latter was replaced by an arbitrary *triangular norm* in the sense of Menger [275], namely an associative operation $*$ on $\mathbb{I}$ satisfying the condition that $(\mathbb{I}, \vee, *)$ is a semiring. Similarly we have the notion of a *triangular conorm* on $\mathbb{I}$, namely an associative operation $*$ on $\mathbb{I}$ satisfying the condition that $(\mathbb{I}, \wedge, *)$ is a semiring. In [276], Menger interpreted triangular norms in the context of continuum physics as rules for generating new probabiliistically-determined objects from existing ones in the psychophysical continuum space. In [55], Butnariu and Klement use triangular norms for interaction rules of economic agents in fuzzy games.

Menger's work was extended by Wald [393] and others. For a summary of the development of this notion, refer to [356]. Triangular norms have also been extensively applied to image processing, optimization, and many other areas of applied mathematics and computer science. See [55, 280] for an extensive study of triangular norms. Triangular norm-based propositional fuzzy logics are studied in [56]. Such logics include the usual min-max logic as well as Lukasiewicz logic. An extensive literature exists on the identification of many infinite families of such norms and their suitability for various real-life applications. For background information on multi-valued logics, refer to [338].

A special case of Zadeh's work is the *interval analysis* of Moore [287], which was initially developed to handle computation with numbers the exact values of which have become uncertain due to repeated roundoff and truncation errors in the computation process.

On the other hand, Goguen [138] extended Zadeh's construction by replacing $\mathbb{I}$ by an arbitrary bounded distributive lattice, a lead which has been followed by many others. If R is taken to be a finite totally-ordered set, we get the theory of *f-sets* developed by Toth [379]. These form a topos, unlike the situation with fuzzy sets [231]. In another generalization of multisets, R is taken to be the set of all cardinal numbers. For an axiomatization of this theory, refer to [240].

(IV) EXAMPLE. Dubois and Prade [89] went in a different direction. Given a nonempty set A, a *toll subset* of A is defined by a *cost of membership function* in R^A, where $R = \mathbb{R}^+ \cup \{\infty\}$ is the set of nonnegative real numbers to which ∞ has been adjoined. Toll subsets are used extensively to study shortest-path problems in graph theory. See [154] for details. Again, there are several ways of defining a semiring structure on $\mathbb{R}^+ \cup \{\infty\}$. It is generally accepted that addition should be given by $r_1 \wedge r_2 = min\{r_1, r_2\}$ and so such semirings are additively-idempotent. Again, there are infinitely-many possible ways of defining multiplications on $\mathbb{R}^+ \cup \{\infty\}$ in order to turn it into a semiring. The most common of these, as we have already noted, is just ordinary addition. Any such operation of multiplication is called a *triangular conorm* on $\mathbb{R}^+ \cup \{\infty\}$. Refer to [277, 284] for more details. A first-order theory to describe elements of R^A, considered as "real-valued multisets", is given in [43]. For the relation of this semiring to *penalty theory* in artificial intelligence, refer to [94].

(V) EXAMPLE. If A and B are nonempty sets and $(\mathbb{P}(B), \cup, \cap)$ is the semiring of all subsets of B, then the elements of $\mathbb{P}(B)^A$ are the *multivalued functions* from A to B. Such functions, of course, correspond bijectively to relations on $A \times B$, but the study of structures of the form $\mathbb{P}(\mathbb{R})^A$, per se, has a long history in analysis. For applications of structures of the form $\mathbb{P}(B)^A$ to topology, see [108]. Functions in $\mathbb{P}(B)^A$, where B is a finite set, are now playing an important part in the design of parallel computers. Functions belonging to $\mathbb{P}(A)^{A \times A}$ for some nonempty set A are called *power operations* in [52], where a general theory of such operations is constructed.

If B_0 is a subset of B we can consider the semiring $(\mathbb{P}(B, B_0), \cup, \cap)$ of all subsets of B containing B_0 (note that this is not a subsemiring of $(\mathbb{P}(B), \cup, \cap)$ but rather a homomorphic image of it under the morphism of semirings $B' \mapsto B' \cup B_0$) and look at semirings of the form $\mathbb{P}(B, B_0)^A$. Such constructions appear, for example, in the theory of *supervisors* for automata, developed in [119].

Another theory designed for characterizing situations under uncertainty is Dempster-Shafer theory [78, 360], which is based on certain functions, called *basic assignments*, in $\mathbb{I}^{\mathbb{P}(E)}$, where E is the universal set under consideration. In particular, this theory has proven very useful in dealing with conditions of uncertainty in expert systems. Refer also to [368]. A variant on this appears in [304]: a *complete class of subsets* U of a set E is a collection of subsets of E containing $\varnothing$ and closed under set-theoretic complementation. If U is a complete class of subsets of E then a function $f \in \mathbb{I}^U$ is a *pseudomeasure* or *Sugeno measure* on the space (E, U) if and only if

(1) $f(\varnothing) = 0$ and $f(E) = 1$; and
(2) $A \subseteq B$ in U implies $f(A) \leq f(B)$.

The set of all pseudomeasures on (E, U) is a subsemiring of $\mathbb{I}^{\mathbb{P}(E)}$. A pseudomeasure f on $(E, \mathbb{P}(E))$ satisfying the additional condition that $f(\cup_{i \in \Omega} A_i) = \vee_{i \in \Omega} f(A_i)$ for any nonempty family $\{A_i \mid i \in \Omega\}$ of subsets of E is called a *possibility measure* on E. See [242] and [396]. For a similar theory, refer to [129].

These examples suggest that we consider the general case of structures of the form R^A, where A is a nonempty set and R has a suitable algebraic structure which would include all of the above examples as special cases. The most promising such structure is that of a semiring. A *semiring* is a set R on which we have defined operations of "addition" (usually denoted by +) and "multiplication" (usually denoted by · or by concatenation) such that the following conditions are satisfied:

(1) $(R, +)$ is a commutative monoid with identity element 0_R;
(2) $(R, \cdot)$ is a monoid with identity element 1_R;
(3) Multiplication distributes over addition from either side;
(4) $0_R r = 0_R = r 0_R$ for all $r \in R$;
(5) $0_R \neq 1_R$.

Semirings have been studied for the past 100 years, originating in Dedekind's studies of the algebraic properties of the set of all ideals of a commutative ring and later the work of Vandiver in number theory, but the interest in them grew considerably after Eilenberg used them as a basis for his automata theory in [98]. A general introduction to semiring theory is given in [146] and its extended and revised version, [149]; we will make use of the terminology and notation of those sources, often without explicit comment. Rings and bounded distributive lattices are examples of semirings.

And now the story begins ... .

0. Background Material

As previously mentioned, a *semiring* R is a nonempty set on which we have defined two operations, addition and multiplication, satisfying the following conditions:

(1) $(R, +)$ is a commutative monoid with identity element 0_R;
(2) $(R, \cdot)$ is a monoid with identity element 1_R;
(3) Multiplication distributes over addition from either side;
(4) $0_R r = 0_R = r 0_R$ for all $r \in R$;
(5) $0_R \neq 1_R$.

When the context is unambiguous, we will simply write 0 instead of 0_R and 1 instead of 1_R.

If we do not have a multiplicative identity 1_R, then the structure is called a *hemiring*. It is shown in [146] and [149] that any hemiring can be canonically embedded in a semiring.

Semirings constitute the basic environment in which we will work. The theory of semirings is developed in considerable detail in [146] and [149], of which this monograph is a direct continuation. Notation and terminology will generally follow that source. In this section we emphasize and augment some of the material presented there, as will be needed in the sequel.

Basic semirings. Every semiring R contains a unique minimal subsemiring, namely

$$\mathbb{N} 1_R = \{1_R, 1_R + 1_R, 1_R + 1_R + 1_R, \ldots\}.$$

Following the terminology of [146] and [149], we will call this semiring the *basic semiring* of R and denote it by $R_{(0)}$. All possible basic semirings have been fully characterized by Alarcón and Anderson [17]. Indeed, a basic semiring must either be isomorphic to the semiring $\mathbb{N}$ of all nonnegative integers or to a semiring of the form $B(n, i)$, where $0 \leq i < n$ are positive integers. The underlying set of $B(n, i)$

is $\{0, 1, \ldots, n-1\}$. Addition $\oplus$ in $B(n, i)$ is defined as follows:
  (1) If $0 \leq a + b < n$ then $a \oplus b = a + b$;
  (2) Otherwise, $a \oplus b$ is the unique $0 \leq c < n$ such that $c \equiv a + b \pmod{n-i}$.

This definition uniquely determines the multiplication on $B(n, i)$ as well.

It is straigthforward to check that $B(n, 0) = \mathbb{Z}/n\mathbb{Z}$ for all positive integers n and $B(2, 1) = \mathbb{B}$. Following the terminology of field theory, we define the *characteristic* of a semiring R to be 0 if and only if $R_{(0)} \cong \mathbb{N}$ and (n, i) if and only if $R_{(0)} \cong B(n, i)$.

Let us look at a particular important class of semirings. A *Gel'fand semiring* is a semiring R satisfying the condition that $a + 1$ has a multiplicative inverse for all $a \in R$. By Proposition 3.40 of [146] we note that a Gel'fand semiring has characteristic 0 or $(2, 1)$. A special case of Gel'fand semirings are the *simple* semirings, namely those semirings R satifying the condition that $a + 1_R = 1_R$ for all $a \in R$. All simple semirings have characteristic $(2, 1)$. An entire theory of commutative simple semirings has been developed in [58] under the name of *incline algebras*. These include, as a special case, bounded distributive lattices. Refer also to [32].

We note that if R is a ring then the semiring $(ideal(R), +, \cdot)$ of all two-sided ideals of R, together with R itself, is simple. If R is any ring then the semiring $(R - fil, \cap, \cdot)$ of all topologizing filters of left ideals of R (the product being the Gabriel product) is a simple semiring which is not commutative [144, 151, 363]. This example has been extended to semirings by considering topologizing filters of left congruences [206].

One small observation is important: since $R_{(0)}$ is the unique minimal subsemiring of a semiring R it is, in particular, contained in the center $C(R)$ of R and so elements of of $R_{(0)}$ commute with all elements of R.

The zeroid. If a is an element of a ring satisfying $a + b = b$ for some element b of the ring then a must be 0. In a semiring there may be many such "local zeroes" and, indeed, every element of the semiring may have this property. The extent to which elements of a semiring have this property is one of the ways one measures its "distance" from being a ring.

If R is a semiring we define a relation $\ll$ on R by setting $a \ll b$ if and only if $a + b = b$. We note that:

  (1) $0 \ll b$ for all $b \in R$;
  (2) $a \ll b$ and $b \ll c$ imply $a \ll c$;
  (3) $a \ll b$ and $b \ll a$ imply that $a = b$;
  (4) If $a \ll b$ and $c \in R$ then $a \ll b + c$;
  (5) If $a \ll b$ and $c \ll d$ then $a + c \ll b + d$;
  (6) if $a \ll c$ and $b \ll c$ then $a + b \ll c$;

(7) If $a \ll b$ and $c \in R$ then $ac \ll bc$ and $ca \ll cb$;

(8) If b has an additive inverse and $a \ll b$ then $a = 0$.

Thus, in particular, for any $a \in R$ the set $\{b \in R \mid a \ll b\}$ is an ideal of the semigroup $(R, +)$. Also, we note that if R is simple then $a \ll 1$ for all $a \in R$. In general, the relation $\ll$ is not a partial order. As we shall soon see, a necessary and sufficient condition for it to be a partial order is that $a + a = a$ for all $a \in R$. Simiarly, this condition suffices in order that $a \ll b$ imply that $a + c \ll b + c$ for all $c \in R$. Thus, any bounded distributive lattice has both of these properties.

(0.1) PROPOSITION. *If R is a simple semiring then $ab \ll a + b$ for all $a, b \in R$.*

PROOF. Since R is simple, we have $a + b + ab = a + (1_R + a)b = a + 1_R b = a + b$ and so $ab \ll a + b$. □

The *zeroid* $Z(R)$ of the semiring R is defined to be the set of all elements $a \in R$ satisfying $a \ll b$ for some $b \in R$. If R is a ring then $Z(R) = \{0\}$. If R is a bounded distributive lattice then $Z(R) = R$.

We now look at a weaker relation. If R is an arbitrary semiring we have a relation $\preceq$ defined on R by setting $a \preceq b$ if and only if there exists an element $c \in R$ such that $a + c = b$. Thus, $a \ll b$ surely implies that $a \preceq b$. Again, we note that

(1) $0 \preceq a$ for all $a \in R$;

(2) $a \preceq a$ for all $a \in R$;

(3) If $a \preceq b$ and $b \preceq c$ then $a \preceq c$;

(4) If $a \preceq b$ and $c \in R$ then $a + c \preceq b + c$;

(5) If $a \preceq b$ and $c \in R$ then $ac \preceq bc$ and $ca \preceq cb$.

Furthermore, Wehrung [402] has shown that

(6) If $a \ll b$ and $b \preceq c$ then $a \ll c$.

The relation $\preceq$ is a preorder but not necessarily a partial order on R. A sufficient condition for it to be a partial order is that $a + b + c = a$ imply $a + b = a$ for all $a, b, c \in R$. If the relation $\preceq$ is a partial order on R, then we say that the semiring is *difference ordered*. The semiring $\mathbb{N}$ of natural numbers has this property. Of course, if $a \preceq b$ then the element $c \in R$ satisfying $a + c = b$ need not be unique. Semirings for which it is unique are important in the development of generalized Petri nets [409]. If $\preceq$ is a linear order on R then we note that if $a \neq b$ then the set $\{c \in R \mid a + c = b\}$ is either empty or a singleton. However, there may be many elements c satisfying $a + c = a$.

Note that if R is a simple semiring then condition (5) implies that $ab \preceq a, b$ for all $a, b \in R$. Moreover, if R is multiplicatively idempotent and if a, b, c are elements of R satisfying $c \preceq a, b$ then then $c = c^2 \preceq cb \preceq ab$. This shows that if R

is a difference-ordered multiplicatively-idempotent simple semiring then $(R, \preceq)$ is a meet-semilattice in which $a \wedge b = ab$.

More generally, if R is a partially-ordered semiring satisfying the condition that $0 \leq a$ for all $a \in R$ then for $a, b, c \in R$ we have $a \leq a+b \leq a+b+c$ and so $a+b+c = a$ implies that $a+b = a$. Thus $\preceq$ is also a partial order on R. Moreover, if $a \preceq b$ then $a \leq b$ for any partial order $\leq$ defined on R satisfying $0 \leq a$ for all $a \in R$. See [195]. In [428] the relation $\preceq$ is studied in a more general context and refered to as the "natural order". In [155], this relation is called the "canonical preorder" on R and the semiring R is called a *dioïd* precisely when $\preceq$ is in fact a partial order. A nonempty subset U of a semiring R is *subtractive* when $a \preceq b \in U$ implies that $a \in U$ (in the terminology of poset theory, this says that U is a *lower set*). Subtractive ideals play a very important role in semiring theory [146, 149].

A semiring R is *zerosumfree* if and only if $a + b = 0$ when and only when $a = b = 0$. In other words, R is zerosumfree when and only when $a \not\preceq 0$ for all $0 \neq a \in R$. In particular, if R is zerosumfree then $1 \not\preceq 0$. If R is difference-ordered then it must be zerosumfree. Indeed, if $a + b = 0$ then $0 \preceq a \preceq a + b = 0$ and so $a = 0$. Similarly $b = 0$.

(0.2) PROPOSITION. *The following conditions on a semiring R are equivalent:*

(1) *If $a \preceq b$ and $b \ll c$ in R then $a \ll c$;*
(2) *If $a \preceq b$ and $b \preceq a$ in R then $a = b$.*

PROOF. Assume (1) and let $a, b \in R$ be elements satisfying $a \preceq b$ and $b \preceq a$. Then there exist elements c and d in R satisfying $a + c = b$ and $b + d = a$. Thus $a + c + d = b + d = a$ so $c + d \ll a$. By (1), this implies that $c \ll a$. Therefore $a = a + c = b$.

Now, conversely, assume (2) and let $a \preceq b$ and $b \ll c$ in R. Then there exists an element d of R satisfying $a + d = b$ so $a + d + c = b + c = c$, proving that $a + c \preceq c$. But clearly $c \preceq a + c$ and so, by (2), $c = a + c$, proving that $a \ll c$. □

If a and b are elements of a semiring R, set $[a, b] = \{r \in R \mid a \preceq r \preceq b\}$. This set of course may be empty. Subsets of R of this form are called *intervals* in R. Denote by $int(R)$ the set of all nonempty intervals in R and define operations $[+]$ and $[\cdot]$ on $int(R)$ by setting

$$[a, b] \, [+] \, [c, d] = [a + c, b + d]$$

and

$$[a, b] \, [\cdot] \, [c, d] = [a \cdot c, b \cdot d]$$

for a, b, c, d. It is easy to see that if $J, J' \in int(R)$ then $J[+]J'$ and $J[\cdot]J'$ again belong to $int(R)$ and that, moreover, $(int(R), [+], [\cdot])$ is again a semiring with additive identity $[0,0]$ and multiplicative identity $[1,1]$. Moreover, the function from R to $int(R)$ given by $r \mapsto [r,r]$ is a monic morphism of semirings. Semirings of this form have important applications in many context. Their study began with Moore's "interval analysis" [287], which studied numerical analysis in the context of $int(\mathbb{R})$. This work, as noted, was one of the precursers of Zadeh's fuzzy set theory. For the use of $int(\mathbb{I})$ in representing imprecise probabilities, refer to [391, 412]. For triangular-norm based operators on $int(\mathbb{R} \cup \{-\infty, \infty\})$, see [279]. For $int(R)$, where R is a complete distributive lattice, see [47].

Additively-Idempotent Semirings. Note that the characteristic of a semiring R is $(2,1)$ if and only if $1_R + 1_R = 1_R$, i.e. if and only if the semiring R is *additively-idempotent*. The class of additively-idempotent semirings is a very important one, as we shall see throughout this work. Among its roots is the theory of "gerbiers" studied in [91]. It includes, for example, the *slopes* in the sense of [57], namely those commutative semirings the additive structure of which is a semilattice. It also includes semirings of the form $(\mathbb{I}, \vee, *)$, where $*$ is a triangular norm on $\mathbb{I}$ and those of the form $(\mathbb{I}, \wedge, *)$, where $*$ is a triangular conorm on $\mathbb{I}$, both of which are extremely important in applications and have been extensively studied. Moreover, it includes the *lattice-ordered semirings*, i.e. those semirings R also having the structure of a lattice $(R, \vee, \wedge)$ such that, for all $a, b \in R$, we have $a + b = a \vee b$ and $ab \leq a \wedge b$. We note in passing that if R is a lattice-ordered semiring with idempotent multiplication then we have $a \wedge b = (a \wedge b)^2 \leq ab$ and hence $a \wedge b = ab$ for all $a, b \in R$. Cunninghame-Green [69] refers to commutative additively-idempotent semirings as *belts*, and uses them extensively in industrial mathematics. If R is simple then surely $1_R + 1_R = 1_R$ and so we see immediately every simple semiring is additively-idempotent.

Note that any additively-idempotent semiring is zerosumfree. Indeed, if R is additively idempotent and if $a + b = 0$ then $a = a + 0 = a + a + b = a + b = 0$ and, similarly, $b = 0$.

Any monoid $(M, *)$ defines an additively-idempotent semiring $(\mathbb{P}(M), \cup, \diamond)$, where

$$A \diamond B = \{a * b \mid a \in A, b \in B\}$$

for all $A, B \in \mathbb{P}(M)$.

The use of arbitrary additively-idempotent semiring in place of $\mathbb{I}$ to model fuzzy logic first appears in [139]. The use of the interval $[-1, 1]$ in place of $\mathbb{I}$ was introduced and justified in [186].

Additively-idempotent semirings also arise naturally in the consideration of *command algebras* [183]. In this situation, the partial order $\preceq$ is the *order of determinacy*: $a \preceq b$ if and only if a is less determinate than b.

A semiring is *entire* if the product of two nonzero elements in it is again nonzero. Entire zerosumfree semirings arise naturally in graph theory and provide considerable information about the structure of graphs. With this in mind, Kuntzmann [225] dubbed them *information algebras*. Refer also to [145] for information about these algebras. Indeed, every finite entire semiring which is not a ring is an information algebra [174]. Also, if R is a semiring which is not a ring then $\{0_R\} \cup \{1_R + r \mid r \in R\}$ is a subsemiring of R which is an information algebra. In the context of fuzzy measure theory, Wang and Klir [406] consider commutative entire semirings of the form $(\mathbb{R}^+ \cup \{\infty\}, +, \cdot)$, where $\mathbb{R}^+$ is the set of all nonnegative real numbers, and where addition and multiplication satisfy the following two additional conditions:

(1) For all elements $a \leq b$ and c of the semiring we have $a + c \leq b + c$ and $ac \leq bc$. (The order here is the usual order on $\mathbb{R}^+ \cup \{\infty\}$.)
(2) If $\lim_{i \to \infty} a_i = a$ and $\lim_{i \to \infty} b_i = b$ then $\lim_{i \to \infty} a_i + b_i = a + b$ and $\lim_{i \to \infty} a_i b_i = ab$.

In [281], the authors show that such a semiring is additively idempotent if and only if $a + b = max\{a, b\}$.

We have already noted that $(\mathbb{I}, \vee, *)$ is an additively-idempotent semiring for any triangular norm $*$ on $\mathbb{I}$. A binary operation $*$ on $\mathbb{I}$ satisfying the criterion that $(\mathbb{I}, \wedge, *)$ is a semiring is a *triangular conorm*. Triangular norms and triangular conorms on $\mathbb{I}$ are both special cases of a more general sort of operation: a binary operation $\diamond$ on $\mathbb{I}$ is an *aggregation* if and only if

(1) $a \diamond 0 = a(1 \diamond 0)$; and
(2) $a \diamond 1 = a[1 - (1 \diamond 0)] + (1 \diamond 0)$

for all $a \in \mathbb{I}$. The triangular norms on $\mathbb{I}$ are then precisely the associative aggregations satisfying $1 \diamond 0 = 0$ and the triangular conorms on $\mathbb{I}$ are precisely the associative aggregations satisfying $1 \diamond 0 = 1$ [273].

If R is an additively-idempotent semiring then the basic semiring of R is $\{0_R, 1_R\}$ and this, of course, is isomorphic to $\mathbb{B}$. Therefore, without loss of generality, we can assume that $\mathbb{B} \subseteq R$ for any additively-idempotent semiring R.

Observe that every additively-idempotent semiring the relation $\ll$ and the relation $\preceq$ coincide. Indeed, if $a \preceq b$ then there exists an element c such that $b = a + c$ and so $a + b = a + a + c = a + c = b$ so $a \ll b$. Moreover, this relation is a partial order, and $Z(R) = R$. In fact, this is the only partial order relation definable on R. Thus, if R is additively-idempotent, the monoid $(R, +)$ is a *divisibility monoid* in the sense of [39] (though written additively, rather than multiplicatively). Also, if $a, b \in R$ then $a + b$ is the least upper bound of $\{a, b\}$ with respect to the partial order $\preceq$ and so $(R, \preceq)$ is a sup-semilattice with bottom element 0. Moreover, we note that if $a \preceq a'$ in R and $b \in R$ then $ab + a'b = (a+a')b = a'b$ so $ab \preceq a'b$. See [195] for details.

We observe that if R is additively idempotent then the equivalent conditions of Proposition 0.2 are satisfied. Indeed, let R be additively idempotent and let $a \preceq b$ and $b \ll c$ in R. Then there exists an element d of R satisfying $a + d = b$ so

$$a + c = a + (b + c) = a + a + d + c = a + d + c = b + c = c,$$

proving that $a \ll c$.

Shubin [362] has classified all of the small additively-idempotent commutative semirings. In particular, he has shown that

(1) There exist three isomorphism classes of additively-idempotent commutative semirings having three elements; in two of these classes the semirings are simple.
(2) There exist 14 isomorphism classes of additively-idempotent commutative semirings having four elements; in seven of these classes the semirings are simple.

Wechler [401] points out the interpretation of additive idempotence as representing nondeterministic choice. This interpretation has importance in theoretical computer science.

Complementations. Let R be a simple semiring. As we have already noted, such semirings are additively idempotent and, on them, the relation $\preceq$ is a partial order satisfying $0 \preceq a \preceq 1$ for all $a \in R$. A *negation* on R is a function $\delta: R \to R$ satisfying $\delta(0) = 1$, $\delta(1) = 0$, and $\delta(b) \preceq \delta(a)$ whenever $a \preceq b$. A negation on R is a *complementation* (sometimes also called a *polarity*) if and only if $\delta^2(a) = a$ for all $a \in R$. This implies, of course, that δ is bijective. A complete distributive lattice with a given complementation is sometimes called a *fuzz* [99, 250].

As an example, we note that the function $\delta: a \mapsto 1 - a$ is a complementation on $(\mathbb{I}, \vee, \wedge)$. This complementation also satisfies the condition that $\delta(a \vee b) \geq \delta(a)\delta(b)$ for all $a, b \in \mathbb{I}$. Such complementations will be important to us later.

Trillis, Alsina, and Valverde [381] give a full characterization of all complementations on any subsemiring R of the semiring $(\mathbb{I}, \vee, \wedge)$ containing 0 and 1: a function $\delta: R \to R$ is a complementation if and only if there is a strictly increasing function $g \in \mathbb{N}^R$ satisfying the following conditions:

(1) $g(0) = 0$;
(2) For each $a \in R$ there exists an element $b \in R$ satisfying $g(a) + g(b) = g(1)$;
(3) $\delta: a \mapsto g^{-1}(g(1) - g(a))$ for all $a \in R$.

Also, the function $a \mapsto a^{-1}$ is a complementation on the semiring $R = (\mathbb{R}^+ \cup \{\infty\}, \wedge, +)$. Mesiar [277] has fully characterized all complementations on this semiring: a function $\delta: R \to R$ is a complementation if and only if there exists a strictly decreasing function $g \in \mathbb{I}^R$ satisfying $g(0) = 1$, $g(\infty) = 0$, and $\delta: a \mapsto g^{-1}(1 - g(a))$ for all $a \in R$.

Complementations in bounded lattices which are isomorphic to direct products of chains are characterized in [171].

Complete Semirings. We now want to define semirings in which we have the possibility of computing infinite sums.

Let R be a semiring and let A be a nonempty set. A family U of functions $f: \Omega \to R$ is *admissible* if and only if to each $f \in U$ we can assign a value $\sum_{i \in \Omega} f(i)$ in R such that the following conditions are satisfied:

(1) If Ω is empty then $\sum_{i \in \Omega} f(i) = 0$;
(2) If $\Omega = \{i_1, \ldots, i_n\}$ then $\sum_{i \in \Omega} f(i) = f(i_1) + \cdots + f(i_n)$;
(3) If $f: \Omega \to R$ belongs to U and $r \in R$ then the function $rf: \Omega \to R$ defined by $(rf): i \mapsto rf(i)$ belongs to U and $\sum_{i \in \Omega}(rf)(i) = r[\sum_{i \in \Omega} f(i)]$;
(4) If $f: \Omega \to R$ belongs to U and $r \in R$ then the function $fr: \Omega \to R$ defined by $(fr): i \mapsto f(i)r$ belongs to U and $\sum_{i \in \Omega}(fr)(i) = [\sum_{i \in \Omega} f(i)]r$;
(5) If $\Omega = \cup_{j \in \Lambda} \Omega_j$ is a partition of Ω then a function $f: \Omega \to R$ belongs to U if and only if the restriction f_j of f to Ω_j belongs to U for each $j \in \Lambda$ and the function $g: \Lambda \to R$ defined by $g: j \mapsto \sum_{i \in \Omega_j} f_j(i)$ belongs to U as well. Moreover, in this situation, $\sum_{i \in \Omega} f(i) = \sum_{j \in \Lambda} g(j)$.

The assignment $f \mapsto \sum_{i \in \Omega} f(i)$ is called a *summation* on the admissible family U.

A *complete semiring* R is a semiring in which the family of all functions with values in R is admissible. That is to say, for every family $\{a_i \mid i \in \Omega\}$ of elements of R we can define an element $\sum_{i \in \Omega} a_i$ of R such that the following conditions hold:

(1) If $\Omega = \emptyset$ then $\sum_{i \in \Omega} a_i = 0$;
(2) If $\Omega = \{1, \ldots, n\}$ then $\sum_{i \in \Omega} a_i = a_1 + \cdots + a_n$;
(3) If $b \in R$ then $b[\sum_{i \in \Omega} a_i] = \sum_{i \in \Omega} ba_i$ and $[\sum_{i \in \Omega} a_i]b = \sum_{i \in \Omega} a_i b$;
(4) If $\Omega = \bigcup_{j \in \Lambda} \Omega_j$ is a partition of Ω into the union of disjoint subsets then $\sum_{i \in \Omega} a_i = \sum_{j \in \Lambda} [\sum_{i \in \Omega_j} a_i]$.

Note that if one thinks of infinite sums as a form of integration, as in [271], then condition (4) is just a variant of Fubini's Theorem.

Complete semirings were studied by Eilenberg [98] and Krob [217], based on ideas originally put forth by Conway [66] for the study of automata.

One of the advantages of working in a complete semiring R is that the *Kleene star* $a^* = \sum_{i=0}^{\infty} a^i$ and the *quasi-inverse* $a^+ = \sum_{i=1}^{\infty} a^i = aa^*$ are defined for each element a of R. For this reason, complete semirings were used by Eilenberg in his development of automata theory. Complete distributive lattices are examples of complete semirings. Thus, in particular, the family of all open subsets of a topological space is a complete semiring in which "addition" is $\cup$ and "multiplication" is $\cap$. If R is a complete semiring then the semiring of formal power series $R\langle\!\langle \Sigma^* \rangle\!\rangle$ over a nonempty set Σ with coefficients in R is a complete semiring, as is each semiring of square matrices over R. The Kleene star operation can also exist in more general contexts. Boffa [46] has shown that if R is an additively-idempotent semiring satisfying the condition that for each $a \in A$ the set of all multiplicatively-idempotent elements e for which $e \succeq 1+a$ is nonempty and has a unique minimal element a^*, then this a^* satisfies all of the "rational identities" of the Kleene star, namely, for all $a, b \in R$ we have:

(1) $(a+b)^* = (a^*b)^*a^*$;
(2) $(ab)^* = 1 + a(ba)^*b$;
(3) $a^{**} = a^*$;
(4) $a^* = (a^n)^*(1 + a + \cdots + a^{n-1})$ for all $n \geq 1$.

If T is a partially-ordered set then a subset I of T is *decreasing* if and only if $t_1 \leq t_2 \in I$ implies $t_1 \in I$. The set $R(T)$ of all decreasing subsets of T (including $\varnothing$) is a complete semiring under the operations of union and intersection. In fact, it is a Heyting algebra. A generalization of fuzzy set theory using $R(T)$ (for a suitable T) in place of $\mathbb{I}$, is studied in [330], where the authors discuss several advantages to this approach, due to the richer available structure.

Several important properties of complete semirings are presented in [146] and [149], of which we note the following:

(1) Every complete semiring has a (necessarily unique) infinite element, namely an element ∞ satisfying $a + \infty = \infty$ for all $a \in R$.
(2) Every complete semiring is zerosumfree.

Refer also to [173]. Thus, no nonzero element of a complete semiring has an additive inverse, which shows how very different complete semirings are from rings. Golan and Wang [153] have shown that every additively-idempotent semiring can be embedded in a complete semiring. We also note that one can fruitfully define the notion of limits of sequences in a semiring. See [195] for details.

It is important to note that the additive structure of an complete semiring does not determine its multiplicative structure. For example, let $(M, *)$ be a monoid and let $\mathbb{P}(M)$ be the set of all subsets of M. If $B, B' \in sub(M)$ set $B * B' = \{b * b' \mid b \in B \text{ and } b' \in B'\}$. Then $(\mathbb{P}(M), \cup, \cap)$ and $(sub(M), \cup, *)$ are both additively-idempotent complete semirings.

A summation on a complete semiring R is *necessary* if and only if for all families $\{a_i \mid i \in \Omega\}$ and $\{b_i \mid i \in \Omega\}$ of elements of R satisfying the condition

(*) Each finite subset Λ of Ω is contained in a finite subset Γ of Ω for which

$$\sum_{i \in \Gamma} a_i = \sum_{i \in \Gamma} b_i$$

we must have $\sum_{i \in \Omega} a_i = \sum_{i \in \Omega} b_i$. This condition is not always satisfied. See [146]. An important consequence of this property is the following: let R be a complete semiring with necessary summation and let $a \in R$ satisfy $a + a = a$. Let Ω be a nonempty set and set $a_i = a$ for all $i \in \Omega$. Then $\sum_{i \in \Omega} a_i = a$. Goldstern [153] has constructed an example to show that this is not necessarily true if we do not assume some other condition beyond completeness. Also refer to [195] for a consideration of this point.

(0.3) PROPOSITION. *Let R be an additively-idempotent complete semiring with necessary summation. Let $a \in R$ and let $\{b_i \mid i \in \Omega\}$ be a nonempty subset of R. Then*

(1) *If $a \preceq b_i$ for all $i \in \Omega$ then $a \preceq \sum_{i \in \Omega} b_i$;*
(2) *If $b_i \preceq a$ for all $i \in \Omega$ then $\sum_{i \in \Omega} b_i \preceq a$.*

PROOF. For each $i \in \Omega$ set $a_i = a$. By necessary summation we then have $\sum_{i \in \Omega} a_i = a$.

(1) We are given $a + \sum_{i \in \Omega} b_i = \sum_{i \in \Omega} a_i + \sum_{i \in \Omega} b_i = \sum_{i \in \Omega}(a_i + b_i) = \sum_{i \in \Omega} b_i$ and so $a \preceq \sum_{i \in \Omega} b_i$.

(2) We are given $\sum_{i \in \Omega} b_i + a = \sum_{i \in \Omega} b_i + \sum_{i \in \Omega} a_i = \sum_{i \in \Omega}(b_i + a_i) = \sum_{i \in \Omega} a_i = a$ and so $\sum_{i \in \Omega} b_i \preceq a$. □

A semiring R is a *complete-lattice-ordered semiring (CLO-semiring)* if and only if R has the structure of a complete lattice satisfying the conditions that $a+b = a \vee b$ and $ab \leq a \wedge b$ for all $a, b \in R$. One of the best-known of these semirings is the semiring $(ideal(R), +, \cdot)$ of all ideals of a commutative ring R. CLO-semirings are additively idempotent and have necessary summation. They are also simple and satisfy $\infty = 1$. Note that it is not necessarily true that multiplication distributes over arbitrary joins in a CLO-semiring.

We note that if R has an infinite element – and in particular if R is a complete semiring – then $\{0, \infty\}$ is a subsemiring of R which is canonically isomorphic to $\mathbb{B}$. This is the basic subsemiring of R if and only if R is simple.

We note that CLO-semirings have important applications in several branches of pure and applied mathematics. Some of these deserve special mention:

(1) The *tropical semiring* $(\mathbb{N} \cup \{\infty\}, \wedge, +)$ is a simple semiring that has been used extensively by Imre Simon [364, 365] and others to study the complexity of finite automata. In applications involving the behavior of timed networks, it is known as the *counter time scale*. For a good survey on these semirings, see [320]. For a study of automata over the tropical semiring, also see [210]. It is sometimes necessary to extend this semiring to a

slightly larger one, $(\mathbb{N} \cup \{\omega, \infty\}, \wedge, +)$, in which $i < \omega < \infty$ for all $i \in \mathbb{N}$ and $\omega + u = u + \omega = max\{\omega, u\}$ for all $u \in \mathbb{N} \cup \{\omega, \infty\}$. Refer to [366].

(2) We have already noted the semiring $(\mathbb{R}^+ \cup \{\infty\}, \vee, \wedge)$ in connection with toll sets. Another semiring on the same underlying set, namely the simple semiring $(\mathbb{R}^+ \cup \{\infty\}, \wedge, +)$ has important applications in analysis, as has been emphasized in the papers collected in [272]. Also see [213]. This semiring can also fruitfully replace the semiring $(\mathbb{R}^+, +, \cdot)$ to obtain a new form of probability theory first studied by Maslov (see [271], for example) and later by Akian and her collaborators [14, 16]. This semiring and generalizations thereof also play an important part in multicriteria optimization, optimal control, and the theory of semantic domains, used in the study of the denotational semantics of higher programming languages. Refer, for example, to [371]. For the use of this semiring in the study of timed Petri nets, see [119, 408]. In such situations it is known as the *continuous time scale*. Indeed, so useful is this semiring that recently its arithmetic has been implemented using virtual hardware [25].

(3) The *schedule algebra* $(\mathbb{R} \cup \{-\infty\}, \vee, +)$ (now often called the $(max, +)$-*algebra*) has its origins in optimization theory [68, 132, 133, 212] and has also been used in finding critical paths in graphs [59, 69], in the study of discrete event dynamical systems (see, for example, [26, 119, 120, 122, 164, 302, 303]), control theory [259], and statistical physics. This is a linearly-ordered semiring which is not complete and so not a CLO-semiring.

(4) The *Mascle semiring* $(\mathbb{N} \cup \{-\infty, \infty\}, \vee, +)$ has also been used extensively. See [269, 320]. For applications of $(\mathbb{R} \cup \{-\infty, \infty\}, \vee, +)$, see [119, 405]. In both of these semirings we have the convention that $-\infty + \infty = -\infty$.

For the use of CLO-semirings in defining constraint systems, refer to [40].

If R is a complete semiring and if t is an infinite cardinal then we have a function $\theta_t \colon R \to R$ which assigns to each element $a \in R$ the sum $\theta_t(a)$ of t copies of a. Note that if $a, b \in R$ then $\theta_t(a+b) = \theta_t(a) + \theta_t(b)$ and $\theta_t(ab) = \theta_t(a)\theta_t(b)$. The function θ_t is not a morphism of semirings since, in general, $\theta_t(1_R) \neq 1_R$. However, it is a morphism of hemirings and $im(\theta_t)$ is an additively-idempotent complete semiring.

If R is a CLO-semiring for which multiplication in R distributes over arbitrary joins from either side, then R is called a *quantic-lattice-ordered semiring (QLO-semiring)*. The lattice of all ideals of a ring is a canonical example of a QLO-semiring.

Let X be a nonempty set. The family $\mathbb{P}(X)$ of all subsets of X has the structure of a QLO-semiring, with "addition" defined by $Y_1 + Y_2 = Y_1 \cup Y_2$ and "multiplication" defined by $Y_1 \cdot Y_2 = Y_1 \cap Y_2$. A topology on X is just a subfamily of $\mathbb{P}(X)$ containing $\varnothing$ (the additive identity) and X (the multiplicative identity) and closed under taking arbitrary sums and finite intersections. In other words, the topologies on X are precisely the complete subsemirings of $\mathbb{P}(X)$. Let us look at this a bit differently: it is clear that $\mathbb{P}(X)$ is canonically isomorphic (as a complete semiring) to the product semiring $\mathbb{B}^X$ with componentwise addition and multiplication. Moreover, the set X itself corresponds to the set of all functions $f \in \mathbb{B}^X$ satisfying the condition that $f(x) = 1$ for precisely one element $x \in X$. These functions are precisely the indecomposable idempotent elements of $\mathbb{B}^X$. Finally, we notice that the topologies on X are precisely the complete subsemirings S of $\mathbb{B}^X$.

This way of looking at things suggests that we might replace $\mathbb{B}$ by some other appropriate QLO-semiring. For example, we might want to use $(\mathbb{I}, \vee, *)$, where $*$ is an appropriate triangular norm on $\mathbb{I}$, and thus obtain *fuzzy topologies* on X; we may want to use $(\mathbb{R}^+, \wedge, *)$, where $*$ is an appropriate operation (called a triangular conorm) on $\mathbb{R}^+$ which makes $(\mathbb{R}^+, \wedge, *)$ a semiring, and thus obtain *toll topologies* on X. Or we can take R to be a complete chain or even an arbitrary frame. For examples of such approaches, see [116, 138, 253]. In general, we can consider R^X, where R is a complete semiring, and complete subsemirings S of R^X. See [185] for a general approach to this problem.

See [62] for a study of operations $*$ on $\mathbb{I}$ satisfying the condition that $(\mathbb{I}, \vee, *)$ is a QLO-semiring.

A CLO-semiring R is *frame ordered* if R, as a lattice, is a frame. As was pointed out in [146] and [149], in frame-ordered semirings we can define not only infinite sums but infinite products as well, using a construction based on that in [244]. Let R be a frame-ordered semiring and let $\theta \colon \Omega \to R$. Without loss of generality we can assume that there exists an ordinal h such that Ω is the set of all ordinals less than h. Then we can define the element $a = \prod^r \theta(\Omega)$ inductively as follows:

(1) If $h = 0$ then $a = 1$;
(2) If $h = k + 1 > 0$ is not a limit ordinal and if $\Omega' = \Omega \setminus \{k\}$ then $a = [\prod^r \theta(\Omega')]\theta(k)$;
(3) If $h > 0$ is a limit ordinal then $a = \vee \{\prod^r \theta(\Omega') \mid \Omega' \subset \Omega\}$.

Note that, since R is lattice-ordered, we have $\prod^r \theta(\Omega') \subseteq \prod^r \theta(\Omega)$ whenever $\Omega' \subseteq \Omega$.

We can similarly define $a = \prod^l \theta(\Omega)$ by changing the definition in (2) to be $a = \theta(k)[\prod^l \theta(\Omega')]$. For infinite applications of triangular norms and conorms, see [156].

If A is a nonempty set then the semiring $R = (\mathbb{P}(A^*), +, \cdot)$ and the semiring $R' = (\mathbb{P}(A^\infty), +, \cdot)$ of formal languages and formal ∞-languages on A are QLO-semirings. In the semiring R' we can also define countably-infinite products as follows: if $L_1, L_2, \ldots$ are elements of R'. define $L_1 L_2 L_3 \cdot \ldots$ to be the set of all words $w \in A^\infty$ of the form $w = a_1 a_2 a_3 \cdots$ where, for each i, we have $\square \neq a_i \in (L_i \cap A^*) \cup (L_i \cap A^*)^* \cdot (L_i \cap A^\infty)$.

Residuals. If a and b are elements of a CLO-semiring R we define the *left residual*
$$ab^{\langle -1 \rangle} = \sum \{r \in R \mid rb \leq a\}$$
and the *right residual*
$$b^{\langle -1 \rangle} a = \sum \{r \in R \mid br \leq a\}.$$

Clearly $b^{\langle -1 \rangle} a \wedge ab^{\langle -1 \rangle} \geq a$. Residuals are best studied in the context of QLO-semilattices. See [146] and [149] for details. For a general study of residuation theory refer to [45]. Also refer to [383]. If b has an inverse in R then $(ab^{-1})b \leq a$ and so $ab^{-1} \leq ab^{\langle -1 \rangle}$. On the other hand, if $rb \leq a$ then $r \leq ab^{-1}$ and so $ab^{\langle -1 \rangle} \leq ab^{-1}$. Thus $ab^{\langle -1 \rangle} = ab^{-1}$ and, similarly, $b^{\langle -1 \rangle} a = b^{-1} a$.

The residuals in semirings of the form $(\mathbb{I}, \vee, *)$, where $*$ is some triangular norm, have been studied by Pedrycz [314] under the name of Φ-*operators*. In particular, if $a * b = a \wedge b$, then
$$ab^{\langle -1 \rangle} = \begin{cases} 1 & \text{if } b \leq a \\ a & \text{otherwise.} \end{cases}$$
while if $a * b = max\{0, a + b - 1\}$ then $ab^{\langle -1 \rangle} = min\{1, 1 - a + b\}$. This is also known as *Lukasiewicz-type* residuation. Refer to [383, 399] for other residuals of this form.

In the case that the triangular norm $*$ is continuous, then we in fact have $a \wedge b = ba^{\langle -1 \rangle} * a$ and $a * b^{\langle -1 \rangle} \vee b * a^{\langle -1 \rangle} = 1$ for all $a, b \in \mathbb{I}$. Refer to [385].

In a more general context, Chang [61] defined a *multiple-valued algebra* to be a set R on which we have an operation of addition defined, together with a unitary operation $*$, such that the following axioms are satisfied:

(1) $(R, +)$ is an abelian monoid with neutral element 0;
(2) $r + 0^* = 0^*$ for all $r \in R$;
(3) $r^{**} = r$ for all $r \in R$;
(4) $(r^* + s)^* + s = (s^* + r)^* + r$ for all $r, s \in R$.

Set $1 = 0^*$ and define multiplication on R by $rs = (r^* + s^*)^*$. Set $r \vee s = (rs^*) + s$ and $r \wedge s = (r + s^*)s$ for all $r, s \in R$. Then $(R, \vee, \wedge)$ is a bounded distributive lattice in which residuals are defined by $r^{\langle -1 \rangle} s = r^* + s$.

We note that if R is a CLO-semiring and if $a \leq b$ in R then $0_R b^{\langle -1 \rangle} \leq 0_R a^{\langle -1 \rangle}$. Moreover, $0_R 1_R^{\langle -1 \rangle} = 0_R$ while $0_R 0_R^{\langle -1 \rangle} = 1_R$. Therefore the function $\delta: R \to R$ given by $a \mapsto 0_R a^{\langle -1 \rangle}$ is a negation on R.

The importance of residuals is that, in many contexts, they play the role of implication operators, since CLO-semirings with complementation can be considered as generalizations of Heyting algebras. See [109] for a typical example. Indeed, if R is a CLO-semiring then the function $\lambda\colon R \times R \to R$ given by $(a,b) \mapsto ba^{(-1)}$ or $(a,b) \mapsto a^{(-1)}b$ satisfies the boundary conditions $\lambda(0,0) = \lambda(0,\infty) = \lambda(\infty,\infty) = \infty$ and $\lambda(\infty,0) = 0$. However, refer also to [381] for other alternatives in the case $R = \mathbb{I}$. CLO semirings with complementations are also the proper context in which to consider generalizations of topological spaces. Refer to [185].

Another important application of residuals is to approximate solutions of equations of the form $aX = b$ over suitable semirings – such as matrix semirings over CLO-semirings – where classical methods do not work. Refer to [120, 146, 149].

An element r of a complete semiring R is *compact* if and only if any family $\{r_i \mid i \in \Omega\}$ of elements of R satisfies the condition that if $r \ll \sum_{i \in \Omega} r_i$ then $r \ll \sum_{i \in \Lambda} r_i$ for some finite subset Λ of Ω. Thus, for example, if R is a commutative ring then any finitely-generated ideal of R is a compact element of the semiring of all ideals of R.

Semifields. A semiring R is a *division semiring* if and only if every nonzero element of R has a multiplicative inverse. A commutative division semiring is a *semifield*. Semifields were studied systematically in [282]. Also refer to [176]. The most well-known semifields are the semifield $(\mathbb{Q}^+, +, \cdot)$ of all nonnegative rational numbers and the semifield $(\mathbb{R}^+, +, \cdot)$ of all nonnegative real numbers. The complete linearly-ordered semirings $(\mathbb{N} \cup \{\infty\}, \wedge, +)$, $(\mathbb{R} \cup \{\infty\}, \wedge, +)$, and $(\mathbb{R} \cup \{-\infty\}, \vee, +)$ mentioned above are additively-idempotent semifields. If p is a positive real number, then $(\mathbb{R}^+, +_p, \cdot)$ is a semifield, where $a +_p b = (a^p + b^p)^{1/p}$ for all $a, b \in \mathbb{R}^+$. By the Krull-Kaplansky-Jaffard-Ohm Theorem [134] we know that every additively-idempotent semifield is naturally isomorphic to the semifield of finitely-generated fractional ideals of a Bezout domain.

The semiring $\mathbb{B}$ is also a semifield, and it is easy to see that this is the only finite semifield which is not a field. It is also the only multiplicatively-idempotent semifield. Indeed, if R is a multiplicatively-idempotent semifield and if $0 \neq a \in R$ then $a = a1_R = aaa^{-1} = aa^{-1} = 1_R$ and so $R = \mathbb{B}$.

The characteristic of a semifield which is not a field is either 0 or $(2,1)$ since a semifield which is not a field is zerosumfree [282].

If $\{R_i \mid i \in \Omega\}$ is a set of zerosumfree division semirings then the *pseudodirect product* $R' = \bowtie_{i \in \Omega} R_i$ of the R_i has underlying set

$$\{0\} \cup \times_{i \in \Omega}(R_i \setminus \{0_{R_i}\}).$$

Operations between nonzero elements of R' are defined componentwise, and these operations are extended to all of R' by setting $0 + r' = r' + 0 = r'$ and $0r' = r'0 = 0$

of all $r' \in R'$. Then it is easy to verify that R' is again a division semiring. Indeed, if each R_i is a semifield then R' is a semifield as well.

Maslov extended measure theory by allowing measures on a boolean σ-algebra of subsets of some given universal set to take values in an additively-idempotent semiring rather than in the semifield $(\mathbb{R}^+, +, \cdot)$, thus defining the notion of an *idempotent measure*. Akian [13] uses this approach with values in the semifield $(\mathbb{R} \cup \{-\infty\}, \vee, +)$. Refer to [119] for this, and for the following result.

(0.4) PROPOSITION. *If R is an additively-idempotent semifield then:*
(1) *For $a, b \in R$, the set $\{r \in R \mid r \preceq a \text{ and } r \preceq b\}$ has a unique maximal element, which we will denote by $a \wedge b$; and*
(2) *If $a, b, c \in R$ then $(a \wedge b)c = ac \wedge bc$ and $c(a \wedge b) = ca \wedge cb$.*

PROOF. (1) If $a = 0$ or $b = 0$ then surely $a \wedge b = 0$ since additively-idempotent semirings are positive. Hence assume that both a and b are different from 0. In this case, we claim that $a \wedge b = b(a+b)^{-1}a$. Indeed, if $r \in R$ satisfies $r \preceq a$ and $r \preceq b$ then, since R is positive, we have $a^{-1} \preceq r^{-1}$ and $b^{-1} \preceq r^{-1}$. Therefore $a^{-1} + b^{-1} \preceq r^{-1} + r^{-1} = r^{-1}$ and so

$$r \preceq (a^{-1} + b^{-1})^{-1} = b(a+b)^{-1}a.$$

Hence $a \wedge b$ exists and equals $b(a+b)^{-1}a$.

(2) This follows from the fact that $r \preceq ac$ and $r \preceq bc$ if and only if $rc^{-1} \preceq a \wedge b$. □

If F is a semifield we can define addition and multiplication on F^2 as follows: If $(a, b), (a', b') \in F^2$ then $(a, b) + (a', b') = (a + a', b + b')$ and $(a, b) \cdot (a', b') = (aa' + bb', ab' + a'b)$. It is straightforward to verify that, under these definitions, $(F^2, +, \cdot)$ is a semiring with additive identity $(0, 0)$ and multiplicative identity $(1, 0)$. Moreover, the function $a \mapsto (a, 0)$ is a monic semiring homomorphism. If the semiring F is zerosumfree then $(a, b) \cdot (a', b') = (0, 0)$ implies that

$$aa' + bb' = 0 = ab' + a'b$$

and so $(a+b)(a'+b') = 0$ in F. Since semifields are entire, this implies that $a + b = 0$ or $a' + b' = 0$ and so $a = b = 0$ or $a' = b' = 0$. Thus $(F^2, +, \cdot)$ is entire and so is an Øre semiring in the sense of [146, 149] and so can embedded in a semifield of fractions. We also note that the function $(a, b) \mapsto (a, b)^* = (b, a)$ is an involution of F^2 satisfying $(x + y)^* = x^* + y^*$ and $(x \cdot y)^* = x^* \cdot y = x \cdot y^*$ for all $x, y \in F^2$.

Multiplicatively-cancellative semirings. A semiring R is *multiplicatively cancellative* if and only if for all $a, b \in R$ and all $0 \neq c \in R$ we have $a = b$ whenever $ac = bc$ or $ca = cb$. Division semirings – and indeed all subsemirings of division semirings – are clearly multiplicatively cancellative. The semiring $(\mathbb{N}, +, \cdot)$ is multiplicatively cancellative but not a division semiring. Using techniques adapted from ring theory, it is straightforward to show that if R is a multiplicatively-cancellative commutative semiring then R can be embedded in a semifield, called its *semifield of fractions*. In particular, this implies that if R is a multiplicatively-cancellative commutative semiring which is not a ring (and so, in particular, if it is additively idempotent) then R must be zerosumfree.

(0.5) PROPOSITION. *If R is a commutative, additively-idempotent semiring that is multiplicatively-cancellative then $(a+b)^n = a^n + b^n$ for all $a, b \in R$ and all $0 < n \in \mathbb{N}$. Moreover, if $a \neq b$ are distinct elements of R then $a^n \neq b^n$ for all positive integers n.*

PROOF. If $a = 0$ or $b = 0$ the result is trivial, and so we can assume that both of these elements are nonzero. Since R is zerosumfree, this implies that $a + b \neq 0$. The result is surely true for $n = 1$. Assume therefore that $n > 1$ and that we have already established that $(a+b)^{n-1} = a^{n-1} + b^{n-1}$. By additive idempotence it is easily verified that

$$(a+b)^k = \sum_{i=0}^{k} a^i b^{k-i}$$

for all positive integers k. In particular,

$$(a^n + b^n)(a+b)^{n-1} = (a^n + b^n)\left(\sum_{i=0}^{n-1} a^i b^{n-1-i}\right) = \sum_{i=0}^{2n-1} a^i b^{2n-1-i}.$$

On the other hand, by the induction hypothesis we have

$$(a+b)^n (a+b)^{n-1} = (a+b)^n (a^{n-1} + b^{n-1})$$
$$= \left(\sum_{i=0}^{n} a^i b^{n-i}\right)(a^{n-1} + b^{n-1})$$
$$= \sum_{i=0}^{2n-1} a^i b^{2n-1-i}.$$

and so $(a+b)^n = a^n + b^n$ by cancellation. This proves the first contention.

Now assume that a and b are elements of R satisfying $a^n = b^n$ for some positive integer n. We must show that $a = b$. If $n = 1$ we are done, so assume $n > 1$. Since R is additively idempotent, this means that $a^n + b^n = a^n + a^n = a^n$. But $a^n + b^n = (a+b)^n = \sum_{i=0}^{n} a^i b^{n-i}$ and so, in particular, $a^n = a^n + ab^{n-1}$. Thus

$$(a+b)^n = a^n + b^n = a^n + ab^{n-1} = a(a^{n-1} + b^{n-1}) = a(a+b)^{n-1}$$

and so, by multiplicative cancellation, $a + b = a$. A similar proof shows that $a + b = b$ and so $a = b$, as desired. □

In other words, if R is a commutative, additively-idempotent semiring which is multiplicatively-cancellative then the function from R to itself given by $a \mapsto a^n$ is an injective morphism of semirings for each positive integer n. These morphisms are not necessarily surjective. If the morphism $a \mapsto a^n$ is surjective for each positive n then the semiring R is *algebraically closed*.

(0.6) PROPOSITION. *If R is a QLO-semifield then:*
1. $\bigwedge_{i \geq 1} a^i = 0$ *for all $1 \neq a \in R$.*
2. *R is linearly-ordered if and only if meets in R distribute over arbitrary joins.*

PROOF. (1) We first claim that $\bigwedge_{i \geq h} a^i = \bigwedge_{i \geq k} a^i$ for all $h, k \in \mathbb{N}$. This is surely clear if $a = 0$ so assume that $0 \neq a < 1$. Then

$$\bigwedge_{i \geq h} a^i = \bigwedge_{i \geq h+1} a^{-1} a^i = a^{-1} \left(\bigwedge_{i \geq h+1} a^i \right) \leq \bigwedge_{i \geq h+1} a^i$$

and

$$\bigwedge_{i \geq h+1} a^i = a \left(\bigwedge_{i \geq h} a^i \right) \leq \bigwedge_{i \geq h} a^i$$

so we have $\bigwedge_{i \geq h} a^i = \bigwedge_{i \geq h+1} a^i$, which suffices to establish the claim.

(2) If R is linearly-ordered, clearly meets in R distribute over arbitrary joins. Conversely, assume this condition holds. Since R is a QLO-semiring, it is additively idempotent and simple. If $a, b \in R$ we must show that $a \leq b$ or $b \leq a$, i.e. that $a + b = a$ or $a + b = b$. This is trivial if either a or b equals 0 so assume that that is not the case. Set $c = a + b$ and assume that $c \neq a$. Set $a_1 = c^{-1} a$ and $b_1 = c^{-1} b$. Then $a_1 + b_1 = 1$ so $a_1 < 1$.

We claim that $a_1^n + b_1 = 1$ for all positive integers n. This has already been noted for the case $n = 1$. Now assume inductively that we have already shown that $a_1^{n-1} + b_1 = 1$ for some $n > 1$. Then

$$\begin{aligned}
1 &= a_1^{n-1} + b_1 \\
&= (a_1^{n-1} + b_1)(a_1 + b_1) \\
&= a_1^n + a_1 b_1 + a_1^{n-1} b_1 + b_1^2 \\
&= a_1^n + a_1 b_1 + b_1 (a_1^{n-1} + b_1) \\
&= a_1^n + a_1 b_1 + b_1 \\
&= a_1^n + (a_1 + b_1) b_1 \\
&= a_1^n + b_1
\end{aligned}$$

Therefore, by (1), $1 = \bigwedge_{n \geq 1}(a_1^n + b_1) = \bigwedge_{n \geq 1}(a_1^n \vee b_1) = \left(\bigwedge_{n \geq 1} a_1^n\right) \vee b_1 = 0 \vee b_1 = b_1 = c^{-1}b$ and so $b = c$, which is what we needed to show. $\square$

Other semiring constructions. Quotient semirings are defined by congruence relations. A *congruence relation* on a semiring R is an equivalence relation θ satisfying the additional conditions that if $(a, a'), (b, b') \in \theta$ then $(a+b, a'+b') \in \theta$ and $(ab, a'b') \in \theta$. Note that if θ is a congruence relation on R and if M is a subsemigroup of $(R, +)$ or of $(R, \cdot)$ then θ induces a (semigroup) congruence relation on M.

If θ is a congruence relation on a semiring R and if $a \in R$, we write the equivalence class of a with respect to θ by a/θ and denote the set of all such classes by R/θ. Then R/θ is again a semiring in which addition and multiplication are given by $(a/\theta) + (b/\theta) = (a+b)/\theta$ and $(a/\theta) \cdot (b/\theta) = (ab)/\theta$. This is called the *quotient semiring* of R with respect to θ.

A *left Øre set* of elements of a semiring R is a submonoid A of $(R, \cdot)$ satisfying the following conditions:

(1) For each pair $(a, r) \in A \times R$ there exists a pair $(a', r') \in A \times R$ satisfying $a'r = r'a$;

(2) If $ra = r'a$ for some $r, r' \in R$ and $a \in A$ then there exists an element $a' \in A$ satisfying $a'r = a'r'$;

(3) $0 \notin A$.

Each left Øre set A of elements of R defines an equivalence relation $\sim$ on $A \times R$ by setting $(a_1, r_1) \sim (a_2, r_2)$ if and only if there exist elements u and u' of R satisfying $ur = u'r'$ and $ua = u'a \in A$. See [146] and [149] for details. We denote the set $(A \times R)/\sim$ by $A^{-1}R$ and if B is a nonempty subset of R then we denote $\{a^{-1}b \mid b \in B\}$ by $A^{-1}B$. On the set $A^{-1}R$ we define operations of addition and multiplication as follows:

(1) $(a_1^{-1}r_1) + (a_2^{-1}r_2) = (aa_1)^{-1}[ar_1 + rr_2]$, where $r \in R$ and $a \in A$ are elements satisfying $aa_1 = ra_2$;

(2) $(a_1^{-1}r_1)(a_2^{-1}r_2) = (aa_1)^{-1}rr_2$, where $a \in A$ and $r \in R$ are chosen so that $aa_1 \in A$ and $ar_1 = ra_2$.

Then $(A^{-1}R, +, \cdot)$ is again a semiring, called the *classical left semiring of fractions* of R with respect to A. The *classical right semiring of fractions* of R is defined similarly. See [146] and [149] for details and examples.

Note that for every left Øre set A of elements of a semiring R we have a morphism of semirings $\gamma_A \colon R \to A^{-1}R$ defined by $\gamma_a \colon r \mapsto 1^{-1}r$.

Semimodules over semirings. Let R be a semiring. A *left R-semimodule* is a commutative monoid $(M, +)$ with additive identity 0_M for which we have a function $R \times M \to M$, denoted by $(r, m) \mapsto rm$ and called *scalar multiplication*, which satisfies the following conditions for all elements r and r' of R and all elements m and m' of M:

(1) $(rr')m = r(r'm)$;
(2) $r(m + m') = rm + rm'$;
(3) $(r + r')m = rm + r'm$;
(4) $1_R m = m$;
(5) $r0_M = 0_M = 0_R m$.

Right R-semimodules are defined in an analogous manner. The theory of semimodules over semirings is extensively developed in [146, 149]. If m is an element of a R-module M then an element m' of M satisfying $m + m' = 0_M$ is an *additive inverse* of m. Clearly additive inverses, if they exist, are unique, and we will denote the additive inverse of m, if it exists, by $-m$. The set $V(M)$ of all elements of M having additive inverses is nonempty, since $0 \in V(M)$. An R-semimodule M is *zerosumfree* if and only if $V(M) = \{0\}$. At the other extreme, an R-semimodule M satisfying $V(M) = M$ is an *R-module*. A nonempty subset N of a left R-semimodule M is a *subsemimodule* of M if and only if N is closed under addition and scalar multiplication. Note that this implies that $0_M \in N$. One defines *R-homomorphisms* between left R-semimodules in the same manner as they are defined between modules over a ring.

Semimodules over semirings, and even over semifields, have important applications. For example, let $R = (\mathbb{R} \cup \{\infty\}, min, +)$ and consider $M = R^{\mathbb{R}}$ as a left R-semimodule. Elements of M are *signals*. Addition in M corresponds to *parallel composition* of signals, and scalar multiplication corresponds to *amplification* of signals. See [26] for an analysis of this situation and its applications to systems theory and signal processing. It is easy to verify that $M = (\mathbb{R} \cup \{\infty\}, \wedge)$ is a left $\mathbb{R}^+$-semimodule. Every n-tuple $x = (m_1, \ldots, m_n)$ of elements of M defines an $\mathbb{R}^+$-homomorphism $\gamma_x \colon (\mathbb{R}^+)^n \to M$ by

$$\gamma_x \colon (a_1, \ldots, a_n) \mapsto min\{a_i m_i \mid 1 \leq i \leq n\} = min\{m_i \mid a_i > 0\}.$$

This allows us to consider linear optimization problems in the context of homomorphisms of semimodules, as is done in detail in [428]. Another application of semimodule theory to optimization is the following: let R be the semifield $(\mathbb{R} \cup \{\infty\}, \wedge, +)$, on which we have a metric d, defined by $d(a, b) = |e^{-a} - e^{-b}|$. For a locally-compact topological space X, let $C_0(X)$ be the R-semimodule of all continuous functions $f \in R^X$ satisfying the condition for each $\epsilon > 0$ there exists a compact subset K of X such that $d(f(x), \infty) < \epsilon$ for all $x \in X \setminus K$. The study of

R-homomorphisms of the form $C_0(X) \to C_0(Y)$ is significant in the analysis of a wide range of deterministic problems in optimal control theory, and is developed for this purpose in [211].

1. Powers of a semiring

Let R be a semiring and let A be a nonempty set. Then the set R^A of all functions for A to R has the structure of a semiring in which addition and multiplication are defined elementwise:

$$f + g : a \mapsto f(a) + g(a)$$

and

$$fg : a \mapsto f(a)g(a)$$

for all $f, g \in R^A$. Similarly, if R is a CLO-semiring and if $f, g \in R^A$ we define $fg^{\langle-1\rangle}$ and $g^{\langle-1\rangle}f$ by setting

$$fg^{\langle-1\rangle} : a \mapsto f(a)g(a)^{\langle-1\rangle}$$

and

$$g^{\langle-1\rangle}f : a \mapsto g(a)^{\langle-1\rangle}f(a).$$

(This can be extended a bit: if U is a nonempty subset of R^A then $\sum U$ and $\prod U$ can be defined elementwise so long as $\{f \in U \mid f(a) \neq 0\}$ is finite for each $a \in A$.) The set A will be called the *exponent set* of this semiring. The semiring R^A is complete if R is. Also, as a consequence of the above we note that $f \ll g$ in R^A if and only if $f(a) \ll g(a)$ in R for all $a \in A$ and $f \preceq g$ in R^A if and only if $f(a) \preceq g(a)$ in R for all $a \in A$.

If $f \in R^A$ then the *support* of f is $\mathrm{supp}(f) = \{a \in A \mid f(a) \neq 0\}$. It is clear that the set of all functions $f \in R^A$ having finite support, to which we also adjoin the multiplicative identity of R^A, is a subsemiring of R^A, which we will denote by $R^{(A)}$. If A is finite then, of course, we have $R^{(A)} = R^A$. Otherwise, $R^{(A)}$ is an ideal of R^A.

(1.1) PROPOSITION. *If R is a zerosumfree semiring and if A is a nonempty set then $supp(f+g) = supp(f) \cup supp(g)$ for all $f, g \in R^A$.*

PROOF. Let $f, g \in R^A$. If $a \in supp(f+g)$ then $0 \neq f(a) + g(a)$ and so either $f(a) \neq 0$ or $g(a) \neq 0$, which is to say that $a \in supp(f) \cup supp(g)$. Conversely, suppose that $a \in supp(f) \cup supp(g)$. Then either $f(a) \neq 0$ or $g(a) \neq 0$ and so, since R is zerosumfree, $(f+g)(a) = f(a) + g(a) \neq 0$. □

If $\emptyset \neq B \subseteq A$ then any $f \in R^A$ can be written as $f = f_1 + f_2$, where $supp(f_1) \subseteq B$ and $supp(f_2) \subseteq A \setminus B$. Note too that if R is zerosumfree and if $f, g \in R^A$ satisfy $f \preceq g$ then $supp(f) \subseteq supp(g)$.

If $f, g \in R^A$ we let $f \neg g$ denote the element of R^A defined by

$$(f \neg g): a \mapsto \begin{cases} f(a) & \text{if } a \notin supp(g) \\ 0 & \text{otherwise.} \end{cases}$$

If R is additively idempotent then $f \neg g + f = f$ for all $g \in R^A$.

If A is a nonempty set, then every $a \in A$ defines a morphism of semirings $\epsilon_a: R^A \to R$ by $\epsilon_a: f \mapsto f(a)$. This is called the *evaluation morphism at a*. There is also an injective morphism of semirings $\gamma: R \to R^A$ given by $\gamma(r): a \mapsto r$ for all $a \in A$. Since $\epsilon_a \gamma(r) = r$ for all $a \in A$ and $r \in R$, we see that each ϵ_a is surjective.

The set R^A is also, of course, an R-semimodule with scalar multiplication on either side. If R is commutative then $(rf)(r'f') = (rr')(ff')$ for all $r, r' \in R$ and all $f, f' \in R^A$ so it is in fact an R-semialgebra. The image of γ is a subsemiring of R^A isomorphic to R and so we see that $R^A_{(0)} = R_{(0)}$ for any nonempty set A.

Usually, in this book, the set A will have some additional algebraic structure and we will be concerned with the structure on R^A induced by this additional structure on A.

From an applications point of view, it is often convenient to think of A as being the *underlying data structure* with which we are dealing and of R as the *range of values* our data can assume. The elements of R^A are then various instances of data at our disposal. Pratt [323] suggests considering the elements of A as modeling *events* while the elements of R as modeling *actions*. Pavelka [310] considered subsets of R^A as providing the semantics for an R-valued logic on A.

Even when the set A is assumed to have no structure in itself, the semiring R^A may be interesting and useful, depending on R. Let us look at some examples.

(I) EXAMPLE. Suppose R is the boolean semiring $(\mathbb{B}, \vee, \wedge)$. Then, as we have already noted, R^A can be identified with the semiring of all *crisp subsets* of the set A by considering each $f \in R^A$ as a characteristic function or a *membership indicator function*. This semiring is well-known and well-studied. In a seminal paper [358], the semiring $\mathbb{B}^\mathbb{N}$ was used by Scott as a universal domain for modelling data types. The semiring $\mathbb{B}^A$ is, of course, in fact a complete distributive lattice. It is often useful to iterate this construction. For example, consider the following situation, which appears in task resource modeling: if A is a finite set of elements called *tasks* and B is a finite set of elements called *resources* then a function $f \in (\mathbb{B}^B)^A$ represents a method of allocating to each task a subset of the resources necessary for its completion. See [121] for an example of this approach.

(II) EXAMPLE. The general framework for the study of constraint satisfaction problems in optimization was posed in [40] as follows: let R be a finite multiplicatively-idempotent commutative semiring, let V be an ordered set of variables, and let D be a nonempty finite set. A *constraint* is a function from D^B to R, where B is a subset of V. Also refer to [123] for a similar approach.

(III) EXAMPLE. If Ω is a nonempty set then the *free commutative monoid* generated by Ω is the monoid $C(\Omega) = (\mathbb{N}^{(\Omega)}, +)$. For each $i \in \Omega$, let $x_i \in C(\Omega)$ be defined by

$$x_i \colon h \mapsto \begin{cases} 1 & \text{if } h = i \\ 0 & \text{otherwise} \end{cases}.$$

If R is a complete commutative semiring then $R^{C(\Omega)}$ is a free R-algebra in the following sense: given an R-algebra S and a set $\{s_i \mid i \in \Omega\}$ of elements of S, there is a unique morphism of R-algebras $\gamma \colon R^{C(\Omega)} \to S$ satisfying $\gamma(x_i) = s_i$ for all $i \in \Omega$. This fact is used by Lamarche [241] to obtain new models for the lambda calculus.

(IV) EXAMPLE. Suppose that $R = (\mathbb{R} \cup \{\infty\}, \wedge, +)$. As we have seen, this is an additively-idempotent semiring the importance of which in various applications is becoming more and more appreciated and studied. If A is a nonempty set then the semiring R^A is a natural setting for the study of generalizations of operator theory, known collectively as "idempotent analysis". See [272] for details.

(V) EXAMPLE. It is often the case that we look at products of the form R^A, where R is a semiring and A is a partially-ordered set. In this case the elements of R^A are R-valued *pomsubsets* (= *partially-ordered multisubsets*) of A and if A has a total order defined on it then the elements of R^A are R-valued *tomsubsets* (= *totally-ordered multisubsets*) of A. If A is finite and has a total order then the elements of R^A are R-valued *strings* of elements of A. This terminology is based on [323]. Following his terminology further, we say that a set of pomsubsets is an R-valued *process*, just as a set of strings is an R-valued *language*. Pomsubsets are an important model for concurrency. See [31, 137].

(VI) EXAMPLE. Let M be a monoid. The set $\mathbb{B}^{(M)}$ is just the semiring of all finite subsets of the monoid M, together with M itself. Such semirings have important universal properties. For example, let t be a positive integer and let M be the additive monoid $\mathbb{N}^t$. For each $1 \leq i \leq t$ let m_i be the t-tuple $(h_1, \ldots, h_t) \in M$ defined by the condition that

$$h_j = \begin{cases} 1 & \text{for } j = i, \\ 0 & \text{otherwise} \end{cases}$$

and let $f_i \in \mathbb{B}^M$ be defined by

$$f_i : m \mapsto \begin{cases} 1 & \text{if } m = m_i, \\ 0 & \text{otherwise.} \end{cases}$$

Then if S is any additively-idempotent commutative semiring and if $s_1, s_2, \ldots, s_t \in S$ there exists a unique morphism of semirings $\gamma \colon \mathbb{B}^M \to S$ satisfying $\gamma(f_i) = s_i$ for all $1 \leq i \leq t$.

If $M = (\mathbb{N}, +)$ and if R is an arbitrary semiring, then it is easily seen that $R^{(M)}$ is isomorphic to the polynomial semiring $R[X]$ in a commuting indeterminate X over R.

(VII) EXAMPLE. Semirings of the form $\mathbb{P}(B)^A$, where A and B are nonempty sets, often appear in mathmatics and in various models of real-world situations. Thus, for example, each equivalence relation $\equiv$ on a set A defines a function in $\mathbb{P}(A)^A$ which assigns to each element of A its equivalence class under $\equiv$.

One popular model considers B as a set of *resources* and a function $f \in \mathbb{P}(B)^A$ as an allocation of required resources to each member of our reference set A. A use of this model to consider the computer game TETRIS is given in [121]. If we think of A as a set of organisms and of B as a collection of behaviors or traits of those organisms, then a function $f \in \mathbb{P}(B)^A$ assigns to each organism the collection of those behaviors or traits which it has been observed to exhibit. An important subsemiring of this semiring consists of all those functions $f \in \mathbb{P}(B)^A$ satisfying the condition that $f(a)$ is finite for all $a \in A$.

One situation which appears very often is the case in which A and B are metric spaces. If $f \in \mathbb{P}(B)^A$ and $a \in A$ then $f(a) \subseteq B$ is sometimes refered to as the *f-graph* of a in B. A situation of particular interest in functional analysis occurs when B is a metric space and we restrict our consideration to functions f with values in the subsemiring of $\mathbb{P}(B)$ consisting of all (closed) compact subsets.

(VIII) EXAMPLE. Let R be a simple semiring on which we have defined a complementation δ. Let A be a nonempty set. Then δ induces a function $\bar{\delta} \colon R^A \to R^A$ given by $\bar{\delta}(f) \colon a \mapsto \delta(f(a))$ for all $a \in A$ and all $f \in R^A$.

Following the terminology of [55], we say that a subset T of R^A is a *clan* if and only if it contains the constant function $a \mapsto 0$ and is closed under products and the action of $\bar{\delta}$. Thus, for example, if $R = (\mathbb{I}, \vee, *)$, where $*$ is a continuous or measurable triangular norm on $\mathbb{I}$, if $\delta \colon a \mapsto 1 - a$, and if A is a topological or measurable space, then the family of all continuous or measurable functions in R^A is a clan.

If R is a frame-ordered simple semiring with complementation δ and if T is a clan which is also closed under taking countably-infinite products, then T is a *tribe*. When $R = (\mathbb{I}, \vee, *)$ for some triangular norm $*$, tribes in R^A play a crucial role in the construction of fuzzy games, and are extensively discussed in [55]. For example, if A is a nonempty set and if $\mathcal{U}$ is a σ-algebra of subsets of A then $\{\chi_B \mid B \in \mathcal{U}\}$ is a tribe in R^A for each triangular norm $*$. Similarly, if $B \subset A$ then $\{f \in R^A \mid f(a) = 0 \text{ or } f(a) = 1 \text{ for all } a \in A \setminus B\}$ is a tribe in R^A for each triangular norm $*$.

The set of all functions $f \in R^A$ which are either constant or the image of which is contained in $[\frac{1}{3}, \frac{2}{3}]$ is a tribe if $*$ is defined by $r * r' = r \wedge r'$ or by

$$r * r' = \begin{cases} r \wedge r' & \text{if } r \vee r' = 1 \\ 0 & \text{otherwise} \end{cases}$$

but not for $r * r' = (r + r' - 1) \vee 0$.

Tribes play an important part in defining $\mathbb{I}$-valued measures on sets.

We now consider some properties of powers of a semiring.

(1.2) PROPOSITION. *If R is a simple semiring then so is R^A for any nonempty set A.*

PROOF. Note that the multiplicative identity of R^A is the function $a \mapsto 1_R$. The result then follows immediately from the definition. □

Since multiplication in R^A is defined elementwise, we note that if R is a simple difference-ordered semiring then for all $f, g \in R^A$ we have $fg \preceq f, g$. Moreover, if R is multiplicatively idempotent and if $h \in R^A$ satisfies $h \preceq f, g$ then $h \preceq fg$ so that $(R^A, \preceq)$ is a meet-semilattice.

(1.3) PROPOSITION. *Let R be an additively-idempotent semiring and let $(A, \leq)$ be a nonempty partially-ordered set. Then the set S of all order-preserving functions from A to $(R, \preceq)$ is a subsemiring of R^A.*

PROOF. Since the additive and multiplicative identities of R^A surely belong to S, we are left to show that S is closed under addition and multiplication. Indeed, if $f, g \in S$ and if $a \leq a'$ in A then

$$(f+g)(a) + (f+g)(a') = f(a) + g(a) + f(a') + g(a')$$
$$= f(a') + g(a')$$
$$= (f+g)(a')$$

and so $(f+g)(a) \preceq (f+g)(a')$. Similarly,

$$(fg)(a) + (fg)(a') = f(a)g(a) + f(a')g(a')$$
$$= f(a)g(a) + [f(a) + f(a')][g(a) + g(a')]$$
$$= f(a)g(a) + [f(a)g(a) + f(a)g(a') + f(a')g(a) + f(a')g(a')]$$
$$= f(a)g(a) + f(a)g(a') + f(a')g(a) + f(a')g(a')$$
$$= f(a')g(a')$$
$$= (fg)(a')$$

and so $(fg)(a) \preceq (fg)(a')$, as desired. □

Let R be a semiring and let A and B be nonempty sets. If $f \in R^A$ and $g \in R^B$ then, following [398], we will say that a function $u: A \to B$ is a *morphism* form f to g if and only if $gu \succeq f$ in R^A. In this case we write $u \in Mor(f, g)$.

If R is a complete semiring and A is a nonempty set then we define the *height* of $f \in R^A$ to be $ht(f) = \sum_{a \in A} f(a)$. If $\{f_i \mid i \in \Omega\}$ is a family of elements of R^A then surely

$$ht\left(\sum_{i \in \Omega} f_i\right) = \sum_{i \in \Omega} ht(f_i).$$

If $f, g \in R^A$ then $ht(fg) \preceq ht(f)ht(g)$. If $ht(f) = \infty$ then f is *normal*. Otherwise it is *subnormal*. If there exists a unique element $a_0 \in A$ such that $ht(f) = f(a_0)$, then f is *unimodal*. If R is a commutative frame-ordered semiring then we can also define the *plinth* of $f \in R^A$ to be $pl(f) = \prod_{a \in A} f(a)$. If $\{f_i \mid i \in \Omega\}$ is a family of elements of R^A then surely

$$pl\left(\prod_{i \in \Omega} f_i\right) = \prod_{i \in \Omega} pl(f_i).$$

For heights and plinths in $(\mathbb{I}, \vee, \wedge)$, refer to [75].

If R and S are semirings then an *R-representation* of S is a monic morphism of semirings from S to R^A for some nonempty set A. One is often interested in obtaining representations of given semirings in terms of semirings having a nicer, or better-studied, structure. For example, one of the questions studied in [58] is when a commutative simple semiring S has an R-representation for some linearly-ordered semiring R.

If R is a partially-ordered semiring and if A is a nonempty set then a function $f \in R^A$ is *bounded* if and only if there exists an element r_f of R satisfying $f(a) \leq r_f$ for all $a \in A$. If the semiring R is additively-idempotent then $bd(R^A) = \{f \in R^A \mid f \text{ is bounded}\}$ is a subsemiring of R.

If the semiring R is partially-ordered then so is R^A, when we set $f \leq g$ if and only if $f(a) \leq g(a)$ for all $a \in A$. Similarly, R^A is (complete, quantic) lattice-ordered whenever R is, with the lattice operations being defined componentwise. In particular, if R is complete then R^A has a unique minimal element, namely the function $f: a \mapsto 0$ and a unique maximal element, namely the function $f: a \mapsto \infty$, where ∞ is the unique maximal element of R.

Points with values in a semiring. If B is a subset of A then the *R-valued characteristic function* of B is the function $\chi_B \in R^A$ defined by

$$\chi_B : a \mapsto \begin{cases} 1_R & \text{if } a \in B \\ 0_R & \text{otherwise.} \end{cases}$$

Note that image of χ_B is in fact contained in $R_{(0)}$.

A function $f \in R^A$ the support of which is at most a singleton in A is called an *R-valued point* of A. Note that R-valued points of A are clearly bounded. We will denote the set of all R-valued points of A by $pt(R^A)$. For a CLO-semiring R, these were studied in [297, 373]. Given $a \in A$ and $r \in R$, it will be convenient to use the notation $p_{a,r}$ to denote the fuzzy point in $pt(R^A)$ with support $\{a\}$ satisfying

$$p_{a,r} : a' \mapsto \begin{cases} r & \text{if } a' = a \\ 0 & \text{otherwise} \end{cases}.$$

In other words, $p_{a,r} = r \cdot \chi_{\{a\}}$. In particular, for any $a \in A$ the point $p_{a,0}$ is just the zero-map, which we will denote simply by p_0 if the context allows us to do so without ambiguity. The point p_0 is *trivial*; all other points are *nontrivial*. If $f \in R^A$ then $f = \sum_{a \in A} p_{a,f(a)}$, where this sum always makes sense since for each $a \in A$, the set $\{p_{b,f(b)} \mid p_{b,f(b)}(a) \neq 0\}$ is surely finite.

If $a \in A$ then $H(a) = \{p_{a,r} \mid r \in R\}$ is surely a left R-semimodule of $R^{(A)}$ (and hence of R^A). We also note that $H(a) \cap H(b) = \{p_0\}$ for $a \neq b$ in A. In particular, this implies that $R^{(A)} = \oplus_{a \in A} H(a)$ in the category of left R-semimodules.

In applications, R-valued sets often represent models of situations being considered, whereas R-valued points often represent initial data, namely information obtained prior to or independent of the model constructed. Therefore it is often very important to consider the precise relationships between an R-valued subset

$f \in R^A$ and the sets of points $\{p_{a,f(a)} \mid a \in B\}$ for various selected subsets B of A. This is particularly true if the set A is assumed, for modeling purposes, to have algebraic structure – say it is a group – which forces a certain structure on $f \in R^A$ – say that of an R-valued subgroup. In such cases, there may be discrepencies between observed data and points coming from the model, which have to be accounted for.

This problem leads to the question of whether it is possible to define the notion of membership of an R-valued point of a nonempty set A in an R-valued subset of A. One plausible definition is to say that $p_{a,r}$ is a member of f if and only if $p_{a,r} \preceq f$, i.e. if and only if $p_{a,r} + g = f$ for some $g \in R^A$. Since $p_{a,r}(b) = 0$ for all $b \neq a$, this condition is equivalent to the condition that $r \preceq f(a)$. Thus, if $f \in R^A$, we will set $pt(f) = \{p_{a,r} \in pt(A) \mid r \preceq f(a)\}$. Note that if $p_{a,r} \in pt(f)$ and if $r' \preceq r$ then $p_{a,r'} \in pt(f)$. Note that if $f, g \in R^A$ then $f \preceq g$ if and only if $pt(f) \subseteq pt(g)$.

If the semiring R is complete, will say that $f \in R^A$ is *point complete* if and only if, for any $a \in A$ and any family $\{r_i \mid i \in \Omega\}$ of elements of R, satisfying $r = \sum_{i \in \Omega} r_i$, we have $p_{a,r} \in pt(f)$ whenever $p_{a,r_i} \in pt(f)$ for each $i \in \Omega$. This condition is discussed in [162]. Note that if f is point complete, then, given the subset $pt(f)$ of R^A, we can reconstruct the function f itself, since

$$f: a \mapsto \sum \{r \in R \mid p_{a,r} \in pt(f)\}$$

for each $a \in A$.

A problem with this definition, however, as pointed out in [182], is that even if R is additively idempotent, it may be possible for an R-valued point to be a member of $\sum_{i \in \Omega} f_i$ but not be a member of f_h for each $h \in \Omega$. In other words,

$$\bigcup_{i \in \Omega} pt(f_i) \subseteq pt\left(\sum_{i \in \Omega} f_i\right)$$

but we do not necessarily have equality. This, however, suffices for the function $f \mapsto pt(f)$ from R^A to the semiring of all subsets of R^A to be a *power domain construction* in the sense of [177, 178, 179]. These constructions, having their origins in the work of Dana Scott, are important in various areas of theoretical computer science, ranging from database theory to models for the semantics of nondeterministic programming languages.

Let R be a semiring and let A be a nonempty set. Note that $R^{(A)}$ is a left R-semimodule which is generated over R by the subset $\{p_{a,1} \mid a \in A\}$ of $pt(A)$. A set B of nontrivial points in $pt(R^A)$ is *good* if and only if

(*) $p_{a,r} \in B$ then $p_{a,s} \notin B$ for all $r \neq s \in R$.

Let us denote the set of all good subsets of $pt(R^A)$ by $good(R^A)$. Then the functions
$$f \mapsto B_f = \{p_{a,f(a)} \mid a \in supp(f)\}$$
and
$$B \mapsto f_B, \text{ where } f_B(a) = \begin{cases} r & \text{if } p_{a,r} \in B \\ 0 & \text{otherwise} \end{cases}$$
is a bijective correspondence between R^A and the family of all subsets of $good(R^A)$. See [288] for details.

2. Relations with Values in a Semiring

If R is a semiring and A is a nonempty set then the semiring $R^{A \times A}$ is the semiring of all R-*valued relations* on A. Note that if A is finite, say $A = \{1, \ldots, n\}$, then this is just the semiring of all $n \times n$ matrices over R. Fuzzy relations (i.e. $\mathbb{I}$-valued relations) on sets were one of the reasons for the creation of fuzzy set theory and so have been extensively studied, see [87, 104, 295, 305, 376] for example. For R-valued relations, where R is a CLO-semiring, see [105].

If $g \in R^{A \times A}$ is an R-valued relation on A and if $f \in R^A$ then, following [343], we say that g is an R-*valued relation on* f if and only if $g(a, a') \preceq f(a)f(a')$ for all $a, a' \in A$. If $R = \mathbb{I}$, this condition says that the degree of membership of a pair of elements in g never exceed the degree of membership of each of the elements in f.

In theoretical computer science, semiring-valued relations on a semiring-valued set represent abstract programs, with different semirings representing different kinds of semantics [400, 401]. In particular, if A is a nonempty set and if $r \in R$ we have the R-valued relation e_r on A defined by

$$e_r \colon (a, a') \mapsto \begin{cases} r, & \text{if } a = a' \\ 0_R, & \text{otherwise} \end{cases}.$$

In other words, $e_r = r\chi_D$, where $D = \{(a, a) \mid a \in A\} \subseteq A \times A$. Note that functions in R^A of the form $a' \mapsto e_r(a, a')$ for given $r \in R$ and $a \in A$ belong to $pt(R^A)$ and any element of $pt(R^A)$ is of this form. Similar constructions also arise in the theory of fuzzy relational databases, which has been investigated by several authors.

Another way of looking at things is the following: if A is a nonempty set the Main and Benson [258] consider the elements of $\mathbb{B}^{A \times A}$ to be nondeteriministic programs on the set A of *states*. Here it is understood that if $f \in \mathbb{B}^{A \times A}$ then

$f(a_1, a_2) = 1$ means that the program f may transform a_1 into a_2. Of course, another way to look at $\mathbb{B}^{A \times A}$ is as a directed graph, where an arc exists between nodes a_1 and a_2 precisely when $f(a_1, a_2) = 1$. If we replace $\mathbb{B}$ by $\mathbb{I}$, we get a fuzzy nondeterministic programs on the set A, or alternatively, a fuzzy graph. We could equally well replace it by an arbitrary semiring.

Now let A and B be nonempty sets. If R is a semiring then an *R-valued relation between A and B* is a function $h \in R^{A \times B}$. If $R = \mathbb{I}$ we have, in particular, the notion of a fuzzy relation between sets. Fuzzy relations have been used extensively to handle uncertainty by nonprobabilistic means, beginning with [351]. For a review of the literature on this subject until 1995, see [81, 90]. For use of fuzzy relations to define associations (such as thesauri) in information retrieval, refer to [283].

Note that if $f \in R^{A \times B}$ is an R-valued relation between A and B then we have a corresponding *opposite relation* $f^{op} \in R^{B \times A}$ between B and A defined by $f^{op} : (b, a) \mapsto f(a, b)$.

If R is a semiring and if A and B are nonempty sets then an R-valued relation $h \in R^{A \times B}$ is sometimes called a *Chu space*. Such spaces have been studied intensively by Pratt [324, 325, 326, 327, 328, 329] and his students, with an eye on applications in computer science. In this approach, A is the set of *events* (or *values, locations, variables, points*) and B is the set of *states* ("possible worlds"). The value $f(a, b)$ represents the extent (or complexity) of the event a happening at state b. In particular, if $R = \mathbb{B}$ then $f(a, b) = 1$ if event a has happened at state b and $f(a, b) = 0$ if it has not. This interpretation has been used in [170, 327] to build models of concurrent systems. Chu spaces, however, are not the only attempt to use structures of the form $R^{A \times B}$ to study the behavior of systems. Refer, for example, to [33]. By allowing R to be any appropriate semiring, one can easily obtain fuzzy and relative versions of the results found there.

In another interpretation [239, 325], one views $h \in R^{A \times B}$ as a *game* between two players, where A is the set of strategies available for one player and B is the set of strategies available for the other. In this situation, $h(a, b)$ is the payoff resulting from the choice of strategies a and b respectively.

Continuing the previous approach developed by Main and Benson, if A and B are nonempty sets then the elements of $\mathbb{B}^{(A \times B) \times A}$ are nondeterministic programs on the set A of *states* and the set B of *inputs*. Again, replacing $\mathbb{B}$ by $\mathbb{I}$ gives us fuzzy nondeterministic programs with inputs. Refer to [236].

Let R be a semiring. If A and B are nonempty sets and if $f \in R^A$ and $g \in R^B$ then we have an R-valued relation $f \times g$ between A and B defined by

$$f \times g : (a, b) \mapsto f(a) g(b).$$

Note that if $f \in R^A$ then an R-valued relation h on A is an R-valued relation on f precisely when $h \preceq f \times f$ in $R^{A \times A}$. In the sequel, we will consider the case in which $(A, *)$ is a semigroup. In this situation, we can also assign to each function $f \in R^A$ the function $f_* \in R^{A \times A}$ defined by $f_*:(a, a') \mapsto f(a * a')$. We will then say that f is an R-valued semigroup of A precisely when $f_* \succeq f \times f$.

If $h \in R^{A \times B}$ is an R-valued relation between A and B then we can think of h as a function $h^\sharp: A \to R^B$ defined by $h^\sharp(a): b \mapsto h(a, b)$ for all $a \in A$ and there are occassionally reasons to do so, especially if this function turns out to be monic.

An R-valued relation $f \in R^{A \times B}$ is an *R-valued function from A to B* if and only if, for each $a \in A$, the set $\{b \in B \mid f(a, b) \neq 0_R\}$ is either empty or a singleton. In other words, f is an R-valued function from A to B if and only if, for each $a \in A$, the function $b \mapsto f(a, b)$ in R^B is an R-valued point. For applications of $\mathbb{N} \cup \{-\infty, \infty\}$-valued functions, see [19]. Note that the rather restrictive condition on the definition of an R-valued function are necessary. Were we to define an R-valued function by the conditions that $f(a, b) = f(a, b') \Rightarrow b = b'$ but allow the possibility of $f(a, b) \neq 0_R$ and $f(a, b') \neq 0_R$ for $b \neq b'$ then it would be easy to construct examples of R-valued functions the composition of which is not an R-valued function. See [299] for details.

If R is a complete semiring and if A, B, C are nonempty sets (or if R is an arbitrary semiring and the set B is finite) and if $h \in R^{A \times B}$ and $k \in R^{B \times C}$ are R-valued relations then we can define the R-valued relation $k \circ h \in R^{A \times C}$ by

$$k \circ h: (a, c) \mapsto \sum_{b \in B} h(a, b) k(b, c).$$

In other words, $k \circ h: (a, c) \mapsto ht(h_a k_c)$, where $h_a, k_c \in R^B$ are respectively defined by $h_a: b \mapsto h(a, b)$ and $k_c: b \mapsto k(b, c)$. It is straightforward to show that $\circ$ is associative and distributes over addition from either side. Also, if $h \preceq h'$ in $R^{A \times B}$ or $k \preceq k'$ in $R^{B \times C}$ then $k \circ h \preceq k \circ h'$ and $k \circ h \preceq k' \circ h$. If k and h are R-valued functions then $k \circ h$ is also an R-valued function. Indeed, if $a_0 \in A$ then, if there exists an element $b_0 \in B$ for which $f(a_0, b_0) \neq 0$ then that element must be unique. Similarly, if there exists an element $c_0 \in C$ for which $g(b_0, c_0) \neq 0$ then that element must be unique. On the other hand, if $(k \circ h)(a_0, c_1) \neq 0$ for some $c_1 \in C$, then there exists an element $b_1 \in B$ such that $h(a_0, b_1) k(b_1, c_1) \neq 0$, which, by the uniqueness of b_0 and c_0, implies that $b_1 = b_0$ and $c_1 = c_0$. However, we do note that $h \circ k$ could be the 0-map even if k and h are not. This would not be so if the semiring R is entire, for then we would have $(k \circ h)(a_0, c_0) = h(a_0, b_0) k(b_0, c_0) \neq 0$.

(2.1) PROPOSITION. *Let R be a complete semiring and let A be a nonempty* set. If $h, k \in R^{A \times A}$ are R-valued relations satisfying $h \circ k = k \circ h$ then

$$\sum_{b \in A}\left[\sum_{x \in A} h(a,x)k(x,b)\right]\left[\sum_{y \in A} h(b,y)k(y,c)\right]$$
$$= \sum_{b \in A}\left[\sum_{x \in A} h(a,x)h(x,b)\right]\left[\sum_{y \in A} k(b,y)k(y,c)\right]$$

for all $a, c \in A$.

PROOF. If $a, c \in A$ then

$$\sum_{b \in A}\left[\sum_{x \in A} h(a,x)k(x,b)\right]\left[\sum_{y \in A} h(b,y)k(y,c)\right]$$
$$= \sum_{b \in A} [(h \circ k)(a,b)][(h \circ k)(b,c)]$$
$$= [(h \circ k) \circ (k \circ h)](a,c)$$
$$= [(h \circ h) \circ (k \circ k)](a,c)$$
$$= \sum_{b \in A} [(h \circ h)(a,b)][(k \circ k)(b,c)]$$
$$= \sum_{b \in A}\left[\sum_{x \in A} h(a,x)h(x,b)\right]\left[\sum_{y \in A} k(b,y)k(y,c)\right]$$

as desired. □

If we consider the particular case in which R is a CLO-semiring and A, B, and C are nonempty sets then there are other possible, and useful, ways of defining compositions between relations $h \in R^{A \times B}$ and $k \in R^{B \times C}$. Some of these, along with their applications, were considered in [28, 30] and then extended in [75] for the case of R being $(\mathbb{I}, \vee, \wedge)$. These can be easily extended further. For example, over an arbitrary CLO-semiring we can consider the *Bandler-Kohout compositions*

$$(h \triangleleft_{BK} k) \colon (a,c) \mapsto \bigwedge_{b \in B} h(a,b)k(b,c),$$

$$(h \triangleright_{BK} k) \colon (a,c) \mapsto \bigwedge_{b \in B} k(b,c)h(a,b),$$

and

$$(h \diamond_{BK} k) \colon (a,c) \mapsto (h \triangleleft_{BK} k) \wedge (h \triangleright_{BK} k),$$

which can be applied to medical diagnosis [29].

We can also set

$$(h \triangleleft k)\colon (a,c) \mapsto \left(\sum_{b\in B} h(a,b)\right) \left(\bigwedge_{b\in B} h(a,b)^{(-1)} k(b,c)\right) \left(\sum_{b\in B} k(b,c)\right).$$

The reasons for considering such compositions are detailed in [75].

If A is a locally-finite partially-ordered set and if we restrict ourselves to the set T of functions $f \in R^{A\times A}$ satisfying the condition that $f(a,b) = 0$ unless $a \leq b$ then $(T, +, \circ)$ is a subhemiring of $R^{A\times A}$ called the *R-valued incidence algebra* on A. See [345] for more details.

Compositions can, of course, be iterated. In particular, if $h \in R^{A\times A}$ we can define $h^{\circ k}$ for all $k \geq 0$ by setting $h^{\circ 0} = e_0$ and then setting $h^{\circ k} = h^{\circ(k-1)} \circ h$ for all $k > 0$. Moreover, if R is complete we can further define $h^{\circ *} = \sum_{k=0}^{\infty} h^{\circ k}$ to be the reflexive and transitive closure of h. These definitions lead to the operational semantics of R-valued computations, as studied in [401].

We also note, of course, that if the sets A, B, and C are finite then the elements of $R^{A\times B}$ and $R^{B\times C}$ are just appropriately-sized matrices over R and that the operation $\circ$ is then just the usual matrix multiplication. The problem of decomposing a given R-valued relation $f \in R^{A\times A}$ into $g \circ h$, for suitable g and h in $R^{A\times A}$, has received considerable attention. Thus, Di Nola et al. [83] have solved the problem of writing an $\mathbb{I}$-valued relation $f \in \mathbb{I}^{A\times A}$, where A is a finite set, in the form $g \circ g$. Refer to [390] for more general results. Such decompositions have important application in control theory.

The notion of an R-valued function has been weakened by Chakraborty and Khare [60], who consider a less restrictive condition: an R-valued relation $h \in R^{A\times B}$ is an *R-valued map* if and only if

(1) If $a \in A$ and there exist elements $b \neq b'$ in B such that $h(a,b)$ and $h(a,b')$ are both nonzero, then $h(a,b) \neq h(a,b')$;
(2) If $a \in A$ then there exists a unique $b \in B$ satisfying $h(a,b) = 1$.

Unfortunately, the composition of R-valued maps is not necessarily an R-valued map, even if R is entire. To see this, let $R = (\mathbb{I}, \vee, \wedge)$ and consider the sets $A = \{a_1\}$, $B = \{b_0, b_1, b_2, \ldots\}$, and $C = \{c_1, c_2\}$. Consider the $\mathbb{I}$-valued maps $f \in \mathbb{I}^{A\times B}$ and $g \in \mathbb{I}^{B\times C}$ given by

$$f\colon (a_1, b_i) \mapsto \begin{cases} 1 & \text{if } i = 0 \\ (i-1)/i & \text{otherwise.} \end{cases}$$

and

$$g\colon (b_i, c_j) \mapsto \begin{cases} 1 & \text{if } j = 1 \\ 0 & \text{if } i = 0 \text{ and } j = 2 \\ (i-1)/i & \text{otherwise.} \end{cases}$$

Then $g \circ f$ is not a $\mathbb{I}$-valued map since $(g \circ f)(a_1, c_1) = (g \circ f)(a_1, c_2) = 1$.

Another weakening of the notion of an R-valued function is based on [263]. Let A and B be nonempty sets and let R be a semiring. If $r_0 \in R$ then an R-valued relation $f \in R^{A \times B}$ is an *R-valued function above r_0* if and only if for each $a \in A$ the set $\{b \in B \mid f(a, b) \succ r_0\}$ is either empty or a singleton. Note that the composition two R-valued functions above r_0 need not be an R-valued function above r_0, even for the "nice" case of $R = \mathbb{I}$.

Let $f \in R^{A \times B}$ and $g \in R^{A' \times B'}$ be R-valued relations. A *transform* $(u, v): f \to g$ consists of a pair of functions $u: A \to A'$ and $v: B' \to B$ satisfying

$$f(a, v(b')) = g(u(a), b)$$

for all $a \in A$ and $b' \in B'$. Note that if $(u, v): f \to g$ and $(u', v'): g \to h$ are transforms then $(u'u, vv'): f \to h$ is also a transform. If there exists a transform $(u, v): f \to g$ then we say that f is a *left adjoint* of g and g is a *right adjoint* of f. This notion is, of course, inspired by the corresponding notion in category theory. In the model of concurrent systems proposed by [170], a transform $(u, v): f \to g$ determines a *simulation* of g by f.

If A is a nonempty set then $\circ$ is an associative operation on $R^{A \times A}$ and it is easy to verify that $(R^{A \times A}, +, \circ)$ is a complete semiring the multiplicative identity of which is the relation e_1. Now assume that R is a CLO-semiring and let A, B, and C be nonempty sets. If $g \in R^{A \times B}$, $h \in R^{B \times C}$, and $k \in R^{A \times C}$ then

$$k \circ h^{\langle -1 \rangle} = \sum \{g' \in R^{A \times B} \mid g' \circ h \leq k\}$$

and

$$g^{\langle -1 \rangle} \circ k = \sum \{h' \in R^{B \times C} \mid g \circ h' \leq k\}.$$

In particular, if $h \in R^{A \times A}$ then

$$h \circ h^{\langle -1 \rangle} = \sum \{k \in R^{A \times A} \mid k \circ h \leq h\}$$

and

$$h^{\langle -1 \rangle} \circ h = \sum \{k \in R^{A \times A} \mid h \circ k \leq h\}.$$

These R-valued relations on $A \times A$ are called, respectively, the right and left *traces* of h and have been studied, for the special case of $R = \mathbb{I}$, in [86, 110, 324].

Let A and B be nonempty sets and let $h \in R^{A \times B}$ be an R-valued relation between A and B. If $f \in R^A$ we define $h[f] \in R^B$ by setting

$$h[f]: b \mapsto \sum_{a \in A} f(a) h(a, b)$$

for all $b \in B$ and if $g \in R^B$ we define $h^{-1}[g] \in R^A$ by setting

$$h^{-1}[g]: a \mapsto \sum_{b \in B} h(a,b)g(b)$$

for all $a \in A$. Again, we note immediately that if $f \preceq f'$ in R^A then $h[f] \preceq h[f']$ in R^B while if $g \preceq g'$ in R^B then $h^{-1}[g] \preceq h^{-1}[g']$ in R^A.

Thus, for example, if $A \subseteq B$ and if $h \in R^{A \times B}$ is the inclusion map, defined by

$$h(a,b) = \begin{cases} 1 & \text{if } a = b \\ 0 & \text{otherwise} \end{cases}$$

then, for each $f \in R^A$ we see that $h[f]$ is the extension of f to a function in R^B having the same support, while for each $g \in R^B$, we see that $h^{-1}[g]$ is the restriction of g to A.

As an example of the above, let us look at a construction due to Wiegandt [404]. Let $(A, \leq)$ be a partially-ordered set satisfying the condition that, for each $a \in A$, the set $\{b \in A \mid b \leq a\}$ is finite. Let $*$ be an operation defined on $\{(a,b) \in A \times A \mid b \leq a\}$. Given $f, g \in R^A$, we define the relation $h \in R^{A \times A}$ by setting

$$h(a,b) = \begin{cases} g(b*a) & \text{if } a \leq b \\ 0 & \text{otherwise} \end{cases}.$$

Then

$$h[f]: b \mapsto \sum_{a \in A} f(b)h(a,b) = \sum_{a \leq b} f(b)g(b*a).$$

Following [149, 149], we write $h[f] = f * g$ and it is easily to verify that $(R^A, +, *)$ is a hemiring, called the *Wiegandt convolution algebra* over A. By Proposition 4.11 of [146] we see that in order for $(R^A, +, *)$ to be a semiring it suffices that the following conditions hold for $a, b, c \in A$:

(1) If $a \geq b$ then $a \geq a * b$ and $a * (a * b) = b$;
(2) If $a \geq b \geq c$ then $a * c \geq b * c$ and $(a * c) * (b * c) = a * b$;
(3) If $a > b \geq c$ then $a * c > b * c$.

If the semiring R is complete we can dispense with the requirement that $\{b \in A \mid b \leq a\}$ be finite for every $a \in A$.

As a specific instance of this construction, let R be an additively-idempotent semiring and let G be a closed subgroup of $(\mathbb{R}^n, +)$ for some positive integer n. Then subtraction is an operation on $\{(a,b) \in G \times G \mid b \leq a\}$ which satisfies conditions (1) - (3). Sambourskiĭ and Taraschchan [348] consider the semiring $(R^G, +, \ominus)$, which – as they illustrate – has many applications.

Functions $h \mapsto h[f]$ can be considered as inference schemes in an uncertain environment and as such include the fuzzy implication operators used in designing fuzzy controllers and fuzzy microprocessors [169]. There are several ways of doing this. For example, we can consider the following construction, based [74]: an *implication* on a semiring R is an operation $\triangleright$ on R satisfying the boundary conditions $0 \triangleright 0 = 0 \triangleright 1 = 1 \triangleright 1 = 1$ and $1 \triangleright 0 = 0$. If A and B are nonempty sets and if $f_0 \in R^A$ and $g_0 \in R^B$ are given R-valued subsets of A and B respectively, then each implication $\triangleright$ on R defines an R-valued relation $f_0 \triangleright g_0 \in R^{A \times B}$ by $(f_0 \triangleright g_0): (a,b) \mapsto f_0(a) \triangleright g_0(b)$. The R-valued modus ponens rule then becomes: If $f_0(a)$ then $g_0(b)$ and if $f(a)$ then $(f_0 \triangleright g_0)[f](b)$. Refer also to [74, 113, 180, 418, 419, 420, 421].

Solving relational equations of the form $h[f] = g$ is important in various modeling and other applied problems. Refer to [80, 82, 84, 315, 316, 317, 318, 351, 352, 387]. For an algorithm to find such solutions in the case $R = \mathbb{I}$, see [49, 50, 200].

Straightforward arguments show the following:

(1) The map $f \mapsto h[f]$ is a morphism of left R-semimodules;
(2) If $f_1 \ll f_2$ then $h[f_1] \ll h[f_2]$;
(3) If $f_1 \preceq f_2$ then $h[f_1] \preceq h[f_2]$;
(4) The map $g \mapsto h^{-1}[g]$ is a morphism of right R-semimodules;
(5) If $g_1 \ll g_2$ then $h^{-1}[g_1] \ll h^{-1}[g_2]$;
(6) If $g_1 \preceq g_2$ then $h^{-1}[g_1] \preceq h^{-1}[g_2]$.

If A, B, C are nonempty sets and if $h \in R^{A \times B}$ and $k \in R^{B \times C}$ then $k \circ h \in R^{A \times C}$ and it is straightforward to show that for each $f \in R^A$ we have $(k \circ h)[f] = k[h[f]]$ while for each $g \in R^C$ we have $(k \circ h)^{-1}[g] = h^{-1}[k^{-1}[g]]$.

Equivalence relations with values in a semiring. An R-valued relation h on a nonempty set A is [*strongly*] *transitive* if and only if

$$h(a, a')h(a', a'') \preceq h(a, a'')$$

[resp. $h(a,a')h(a',a'') \ll h(a,a'')$] for all $a, a', a'' \in A$. Note that if the semiring R is additively idempotent, complete and with necessary summation then the transitivity condition is equivalent to the condition that $h \circ h \preceq h$. (If A is finite then, of course, the completeness condition is unnecessary). Transitivity, in this context, is a very natural condition arising originally from probability theory. If we consider the elements of A as events and the value of $h(a, a')$ as a probability of transition from event a to event a' then we certainly would want to assume that the probability of going from a to a'' should be at least equal to that of first going from a to a' and then from a' to a''.

―――――――CHAPTER II―――――――

Following the approach of [157], we say that an R-valued relation h on a nonempty set A is *weakly transitive* if there exists a function $\theta\colon R \to R$ satisfying the conditions

(1) $\theta(r) \succeq r$ for all $r \in R$;
(2) If $r \succeq r'$ in R then $\theta(r) \succeq \theta(r')$

such that $h(a,a')h(a',a'') \succ \theta(r) \Rightarrow h(a,a'') \succ r$ for all $a,a',a'' \in A$.

For example, if $R = (\mathbb{I}, \vee, *)$ for some triangular norm $*$, if $A = \{a,b\}$, and if $h \in \mathbb{I}^{A \times A}$ is defined by

$$h(a,a) = 1/3$$
$$h(a,b) = 1/2$$
$$h(b,a) = h(b,b) = 1$$

then h is not transitive but is weakly transitive, given the function

$$\theta\colon r \mapsto \begin{cases} 2r & \text{if } r \leq 1/3 \\ \frac{1}{2}(r+1) & \text{otherwise} \end{cases}.$$

González and Marin [157] provide a condition on a triangular norm $*$ such that every transitive $(\mathbb{I}, \vee, *)$-valued relation on a nonempty set A is weakly transitive as an $(\mathbb{I}, \vee, \wedge)$-valued relation.

The R-valued relation $h \in R^{A \times A}$ is *reflexive* if and only if $h(a,a) = 1$ for each $a \in A$. Note that reflexivity is equivalent to the condition that $e_1 \preceq h$, where e_1 is the multiplicative identity of the semiring $(R^{A \times A}, +, \circ)$ as defined previously. Note that if $h \in R^{A \times A}$ is reflexive and if $f \in R^{A \times A}$ then

$$(f \circ h)(a,a') = \sum_{b \in A} f(a,b)h(b,a') \succeq f(a,a')h(a',a') = f(a,a')$$

for all $a, a' \in A$ and so $f \circ h \succeq f$. Similarly, $h \circ f \succeq f$.

The R-valued relation $h \in R^{A \times A}$ is *symmetric* if and only if $h(a,a') = h(a',a)$ for all $a, a' \in A$. Again, following [157], we say that h is *weakly symmetric* if and only if there exists a function $\theta\colon R \to R$ as above satisfying the condition $h(a,a') \succ \theta(r) \Rightarrow h(a',a) \succ r$ for all $a, a' \in A$.

For example, if if $R = (\mathbb{I}, \vee, \wedge)$ for some triangular norm $*$, if A is a nonempty set, if $g\colon A \to \mathbb{I}$ is not constant, and if n is a positive integer then the R-valued relation

$$h\colon (a,a') \mapsto \begin{cases} 1 & \text{if } a = a' \\ g(a)g(a')^n & \text{otherwise} \end{cases}.$$

is not symmetric but is weakly symmetric, given the function $\theta\colon r \mapsto r^{1/(n+1)}$.

A [strongly] transitive R-valued relation h on A is a [strong] R-valued equivalence relation on A if and only if it is reflexive and symmetric. See [295, 359]. For example, if R is an arbitrary semiring and A is a nonempty set then any $g \in R^A$ defines a $\mathbb{B}$-equivalence relation $\sim_g$ on A by setting $a \sim_g a'$ if and only if $g(a) = g(a')$.

Note that if $h \in R^{A \times A}$ if an R-valued equivalence relation on a set A and if $a, a' \in A$ then $h(a, a')^2 = h(a, a')h(a', a) \preceq h(a, a) = 1$ and so $r^2 \preceq 1$ for all $r \in im(h)$.

If $R = (\mathbb{I}, \vee, \wedge)$ then the function $h \in R^{\mathbb{N} \times \mathbb{N}}$ defined by

$$h\colon (m, n) \mapsto \begin{cases} 1 & \text{if } m = n \\ \frac{1}{2} & \text{if } m + n \text{ is even} \\ 0 & \text{otherwise} \end{cases}$$

is an R-valued equivalence relation on $\mathbb{N}$.

If $R = (\mathbb{I}, \vee, \cdot)$ and if A is a nonempty set then any function $f \in R^A$ defines an R-valued equivalence relation $h \in R^{A \times A}$ by

$$h\colon (a, b) \mapsto \begin{cases} \frac{f(a) \wedge f(b)}{f(a) \vee f(b)} & \text{if } f(a) \neq f(b) \\ 1 & \text{otherwise} \end{cases}$$

Indeed, if $R = (\mathbb{I}, \vee, *)$, where $*$ is a continuous triangular norm, then the family of all R-valued equivalence relations on a nonempty set A has been characterized by Valverde [387]. For further results, refer to [189].

For a somewhat weaker version of R-equivalence, using quantales, refer to [35].

More generally, let R be a semiring and let A be a nonempty set. A nonempty subset P of R^A is an R-valued partition of A if and only if the following conditions are satisfied:

(1) If $a \in A$ there exists precisely one element of P satisfying $f(a) \succeq 1$ and in this case $f(a) = 1$. We denote this element by u_a.
(2) Every element of P is of the form u_a for some $a \in A$.
(3) If $a, b, c \in A$ then $u_a(c)u_b(c) \preceq u_a(b) = u_b(a)$.

It is easily checked that any R-valued partition P of A defines an R-valued equivalence relation $h_P \in R^{A \times A}$ by setting $h_P\colon (a, b) \mapsto u_a(b)$ for all $a, b \in A$. Conversely, if $h \in R^{A \times A}$ is an R-valued equivalence relation on A then h defines an R-valued partition P_h of A, namely the set of all functions $h_a \in R^A$ given by $h_a\colon b \mapsto h(a, b)$. For the case of $R = \mathbb{I}$, see [306, 355].

(2.2) PROPOSITION. Let R be a simple semiring and let $h \in R^{A \times A}$ be an R-valued equivalence relation on a nonempty set A. Let $\sim$ be the relation on A defined by $a \sim a'$ if and only if $h(a, a') = 1$.

(1) If $\preceq$ is a partial order, or

(2) If h is strongly transitive

then $\sim$ is an equivalence relation on A;

PROOF. It is easy to see that $a \sim a$ for all $a \in R$ and that $a \sim a'$ implies $a' \sim a$. Now assume that $a \sim a'$ and $a \sim a''$. Then in case (1) we have

$$1 = h(a, a')h(a, a'') \preceq h(a, a'') \preceq 1$$

and so $a \sim a''$, while in case (2) we have

$$1 = h(a, a')h(a, a'') \ll h(a, a'') \ll 1$$

so that in either case we see that $\sim$ is an equivalence relation on A. □

Of course, in the above situation, the simplest case is when the relation $\sim$ is trivial, i.e. when $h(a, a') \neq 1$ when $a \neq a'$. In this case, given $a \in A$ we may be interested in finding elements $a' \in A$ such that $h(a, a')$ is maximal with respect to the relation $\preceq$ (if such elements exist). This problem is closely related to the "post-office" problem in computer science and, at least for finite sets A, there exist many methods of constructing algorithms to solve this problem efficiently. For one such, which is readily adaptable to our situation, see [64].

Let R be a semiring and let A be a nonempty set. If $f, g \in R^{A \times A}$ and if $fg \in R^{A \times A}$ is the function defined by $fg: (a, a') \mapsto f(a, a')g(a, a')$ then fg is surely symmetric and reflexive whenever both f and g are. Moreover, if both f and g are transitive and at least one of them is central then for all $a, a', a'' \in A$ we have

$$\begin{aligned}(fg)(a,a')(fg)(a',a'') &= f(a,a')g(a,a')f(a',a'')g(a',a'') \\ &= f(a,a')f(a',a'')g(a,a')g(a',a'') \\ &\preceq f(a,a'')g(a,a'') \\ &= (fg)(a,a'')\end{aligned}$$

and so fg is transitive as well. Thus we see that if f and g are R-valued equivalence relations on A, one of which is central, then fg is also an R-valued equivalence relation on A. In particular, if R is commutative then the set of all R-valued equivalence relations on any nonempty set A is closed under taking products.

Let $R = (\mathbb{R}^+ \cup \{\infty\}, \wedge, +)$ and let A be a nonempty set. Then an *(extended) pseudometric* on A is just an R-valued equivalence relation on A. Such a function is an *(extended) metric* if and only if the relation $\sim$ on A which it defines is trivial. Thus, for example, if A is the set of all continuous functions from $\mathbb{I}$ to $\mathbb{R}$ then we have an R-valued equivalence relation h on A defined by

$$h: (\varphi, \psi) \mapsto \int_0^1 |\varphi(t) - \psi(t)| dt.$$

Another interesting example, with applications to the analysis of computer programs, is the following: Let A be a nonempty set of "states" and let L be a nonempty set which is the "language" in which we make statements about the elements of A. We assume that there is a distinguished subset $\models$ of $A \times L$ and we say that a state $a \in A$ *satisfies* a statement $\lambda \in L$ if and only if $(a, \lambda) \in \models$. In this case we write $a \models \lambda$. If $L' \subseteq L$ then the set of *models* for L') is the set

$$Mod(M') = \{a \in A \mid a \models \lambda \text{ for all } \lambda \in L'\}$$

and if $A' \subseteq A$ then the *theory* of A' is the set

$$Th(A') = \{\lambda \in L \mid a \models \lambda \text{ for all } a \in A'\}.$$

Now assume that L has a special element $\bot$ satisfying $Mod(\{\bot\}) = \varnothing$ and that there is an operation $\vee$ defined on L satisfying $Mod(\{\lambda \vee \lambda'\}) = Mod(\{\lambda\}) \cup Mod(\{\lambda'\})$ for all $\lambda, \lambda' \in L$. Then the function $A' \mapsto Mod(Th(A'))$ is a closure operator on $\mathbb{P}(A)$. Moreover, in case $L = \{\lambda_1, \lambda_2, \ldots\}$ is countable then we have an $(\mathbb{R}^+ \cup \{\infty\}, \wedge, +)$-valued equivalence relation h defined on A as follows:

(1) $h(a, a) = 0$ for all $a \in A$;
(2) If $a \neq a'$ in A then $h(a, a') = \frac{1}{n}$, where $n = min\{k \mid a \models \lambda_k \text{ and } a' \not\models \lambda_k\}$.

These last examples suggest that, for a general semiring R, we can treat R-valued equivalence relations in the same way we treat duals of pseudometrics. Thus, for example, if R is a semiring and A is a nonempty set, we say that R-valued equivalence relations $h, k \in R^{A \times A}$ are *Lipschitz equivalent* if and only if there exists $s_1, s_2 \in R$ satisfying $s_1 h(a, a') \succeq k(a, a')$ and $s_2 k(a, a') \succeq h(a, a')$ for all $a, a' \in A$. It is easily checked that this is in fact an equivalence relation.

(2.3) PROPOSITION. *Let R be a complete difference-ordered semiring. If h is a R-valued equivalence relation on a nonempty set A then $h \circ h = h$.*

PROOF. We have already noted that $h \circ h \preceq h$. The converse follows since h is reflexive, as noted above. $\square$

(2.4) PROPOSITION. *Let R be s complete simple difference-ordered commutative semiring and let A be a nonempty set. If $h, k \in R^{A \times A}$ are R-valued equivalence relations on A satisfying $h \circ k = k \circ h$ then $h \circ k$ is also an R-valued equivalence relation on A.*

Moreover,

(1) $h \circ k \succeq h$ *and* $h \circ k \succeq k$;
(2) *If g is an R-valued equivalence relation on A satisfying $g \succeq h$ and $g \succeq k$ then $h \circ k \succeq g$.*

PROOF. If $a \in A$ then, by simplicity,

$$1_R \succeq (h \circ k)(a,a) = \sum_{b \in A} h(a,b)k(b,a)$$
$$\succeq h(a,a)k(a,a) = 1_R$$

and so $(h \circ k)(a,a) = 1_R$. Thus $h \circ k$ is reflexive.

If $a, b \in A$ then

$$(h \circ k)(a,b) = \sum_{c \in A} h(a,c)k(c,b)$$
$$= \sum_{c \in A} k(b,c)h(c,a)$$
$$= (k \circ h)(b,a) = (h \circ k)(b,a)$$

and so $h \circ k$ is symmetric.

Finally, if $a, a', a'' \in A$ then, by Proposition 2.3.

$$[(h \circ k)(a,a')][(h \circ k)(a',a'')] \preceq [(h \circ k) \circ (h \circ k)](a,a'')$$
$$= [h \circ (k \circ h) \circ k](a,a'')$$
$$= [h \circ (h \circ k) \circ k](a,a'')$$
$$= [(h \circ h) \circ (k \circ k)](a,a'')$$
$$= (h \circ k)(a,a'')$$

and so $h \circ k$ is transitive. Thus $h \circ k$ is also an R-valued equivalence relation on A.

Since h is an R-valued equivalence relation on A we know that $h \circ k \succeq k$ and, similarly, $h \circ k = k \circ h \succeq h$ since k is an R-valued equivalence relation on A. This proves (1). If g is an R-valued equivalence relation on A satisfying $g \succeq h$ and $g \succeq k$ then, by Proposition 2.3, we see that for all $a, a' \in A$ we have

$$(h \circ k)(a,a') = \sum_{b \in A} h(a,b)k(b,a')$$
$$\preceq \sum_{b \in A} g(a,b)g(b,a')$$
$$= g \circ g = g$$

and so we have (2). $\square$

If A is a nonempty set then a *Fréchet closure operator* on A is a function $c \colon \mathbb{P}(A) \to \mathbb{P}(A)$ satisfying the following conditions:

(1) $c(\emptyset) = \emptyset$;
(2) $B \subseteq c(B)$ for all $B \subseteq A$;
(3) $c(B) \cup c(B') \subseteq c(B \cup B')$ for all $B, B' \subseteq A$;
(4) $c(c(B)) = c(B)$ for all $B \subseteq A$.

For example, a *convex geometry* on a set A, in the sense of [215], is a Fréchet closure operator.

If R is a simple semiring and if A is a nonempty set then any R-valued equivalence relation $h \in R^{A \times A}$ defines a Fréchet closure operator $c_h \colon \mathbb{P}(A) \to \mathbb{P}(A)$ by setting

$$c_h(B) = \left\{ a \in A \;\middle|\; \sum_{b \in B} h(a,b) = 1 \right\}$$

where, as usual, the sum of an empty set is taken to be 0.

Coulon, Coulon, and Höhle [67] consider a generalization of the R-valued equivalence relations on a nonempty set A, namely those transitive symmetric R-valued relations h on A for which $h(a,a)h(a,a')h(a',a') = h(a,a')$ for all $a, a', a'' \in A$. Here $h(a,a)$ is to be interpreted as the extent to which a exists and $h(a,a')$ is the extent to which a and a' coincide.

(2.5) PROPOSITION. *Let R be a semiring and let A be a nonempty set. If $h \in R^{A \times A}$ is a strong R-valued equivalence relation on A then the following conditions are equivalent for $a, b \in A$:*

(1) $h(a,c) = h(b,c)$ for all $c \in A$;
(2) $h(a,b) = 1$.

PROOF. Assume (1). Then, in particular, $h(a,b) = h(b,b) = 1$ and so we have (2). Conversely, assume (2). If $c \in A$ then $h(a,c) \gg h(a,b)h(b,c) = h(b,c)$ and similarly $h(b,c) \gg h(a,c)$, proving (1). □

(2.6) PROPOSITION. *Let R be a semiring and let A be a nonempty set. If $h \in R^{A \times A}$ is a strong R-valued equivalence relation on A then:*

(1) $h(a,c)h(b,c) = 0$ for all $c \in A$ if and only if $h(a,b) = 0$;
(2) $h(a,c) = h(b,c)$ for all $c \in A$ if and only if $h(a,b) = 1$.

If R is a difference-ordered semiring these conditions hold for an arbitrary R-valued equivalence relation h.

PROOF. (1) Assume that $h(a,c)h(b,c) = 0$ for all $c \in A$. Then

$$h(a,b) = 1 \cdot h(a,b) = h(a,a)h(a,b) = h(a,a)h(b,a) = 0.$$

Conversely, assume $h(a,b) = 0$. Then for all $c \in A$ we have $0 \ll h(a,c)h(b,c) = h(a,c)h(c,b) \ll h(a,b) = 0$.

(2) Assume $h(a,c) = h(b,c)$ for all $c \in A$. Then, in particular, $1 = h(a,a) = h(b,a) = h(a,b)$. Conversely, assume that $h(a,b) = 1$. Then for any $c \in A$ we have $h(a,c) \gg h(a,b)h(b,c) = h(b,c)$ and similarly $h(b,c) \gg h(a,c)$, proving that $h(a,b) = h(b,c)$.

To prove the last statement, we note that if R is a difference-ordered semiring then $\preceq$ is a partial order and so the above proofs work with $\ll$ replaced by $\preceq$. □

In particular, if $h \in R^{A \times A}$ is an R-valued equivalence relation on a nonempty set A then, for each $a \in A$, then *equivalence class* of a with respect to that relation is the R-valued subset h_a of A defined by $h_a: a' \mapsto h(a, a')$. Propositon 2.6 then says that, in the given situation,

(1) $h_a h_b$ is the 0-map if and only if $h(a, b) = 0$; and
(2) $h_a = h_b$ if and only if $h(a, b) = 1$.

The set of all equivalence classes of A with respect to an R-valued equivalence relation h is just the R-valued partition P_h we defined previously. Note that if we have a canonical surjection from A to P_h given by $a \mapsto h_a$.

Other definitions of equivalence with values in a semiring have been proposed. Gupta and Gupta [166] argue that reflexivity should be replaced by the weaker condition that $h(a, a') \preceq h(a'', a'') \neq 0_R$ for all $a, a', a'' \in A$. Yeh [413] had earlier proposed the even weaker condition that $h(a, a') \preceq h(a, a)$ for all $a, a' \in A$.

On the other hand, Sasaki [353] replaces reflexivity with a stronger condition. An R-valued relation e on a nonempty set A is an *R-valued equality relation* if and only if it satisfies the condition that $e(a, a') = 1$ when and only when $a = a'$. We then note the following examples found there: Let $R = (\mathbb{I}, \vee, *)$, where $a * b = 0 \vee (a + b - 1)$, and let A be a set on which we are given a metric $d': A \times A \to \mathbb{R}^+$. Then the function $h: A \times A \to \mathbb{I}$ defined by $h: (a, b) \mapsto 1 - 2\pi^{-1} arctan(d'(a, b))$ is an R-valued equality relation on A.

Similarly, let A be a nonempty set and let $d \in \mathbb{I}^{A \times A}$ be a metric. Then the function $h' \in \mathbb{I}^{A \times A}$ defined by $h' : (a, a') \mapsto 1 - d(a, a')$ is an $\mathbb{I}$-valued equality relation on A.

Let R be a semiring and let A and B be nonempty sets for which we are given R-valued equality relations h_A and h_B respectively. An R-valued relation $h \in R^{A \times B}$ is (h_A, h_B)-*compatible* if and only if

(1) If $a \in A$ then there exists a $b \in B$ for which $h(a, b) \neq 0$;
(2) If $h(a, b) \neq 0$ and $h(a', b') \neq 0$ then $h_B(b, b') \succeq h_A(a, a')[h(a, b) + h(a', b')]$.

The following result is also based on [353].

(2.7) PROPOSITION. *Let R be an additively-idempotent complete entire semiring with necessary summation. Let A, B, C be nonempty sets on which we have defined R-valued equality relations h_A, h_B, h_C respectively. Let $h \in R^{A \times B}$ be (h_A, h_B)-compatible and let $g \in R^{B \times C}$ be (h_B, h_C)-compatible. Then $g \circ f \in R^{A \times C}$ is (h_A, h_C)-compatible.*

PROOF. If $a \in A$ then there exists a $b \in B$ satisfying $h(a, b) \neq 0$ and there exists a $c \in C$ satisfying $g(b,c) \neq 0$. Since R is entire, we then have $(g \circ h)(a,c) \succeq h(a,b)g(b,c) \neq 0$.

Now assume that $a, a' \in A$ and $c, c' \in C$ satisfy $(g \circ h)(a,c) \neq 0$ and $(g \circ f)(a',c') \neq 0$. We must show that $h_C(c,c') \succeq h_A(a,a')[(g \circ h)(a,c) + (g \circ h)(a',c')]$. And, indeed, since $(g \circ h)(a,c) \neq 0$ there exists a $b \in B$ such that $h(a,b) \neq 0$ and $g(b,c) \neq 0$. Similarly, there exists a $b' \in B$ such that $h(a',b') \neq 0$ and $g(b',c') \neq 0$. Moreover, $h_B(b,b') \succeq h_A(a,a')[h(a,b) + h(a',b')]$ and $h_C(c,c') \succeq h_B(b,b')[g(b,c) + g(b',c')]$ and so

$$h_C(c,c') \succeq (h_A(a,a')[f(a,b) + f(a',b')]) [g(b,c) + g(b',c')]$$
$$= h_A([h(a,b)g(b,c) + h(a',b')g(b,c) + h(a,b)g(b',c') + h(a',b')g(b',c')]$$
$$\succeq h_A(a,a')[h(a,b)g(b,c) + h(a',b')g(b',c')].$$

Therefore, by Proposition 0.3(2), we have

$$h_C(c,c') \succeq \sum_{b,b' \in B} h_A(a,a') [h(a,b)g(b,c) + h(a',b')g(b'c')]$$
$$= h_A(a,a') \left[\sum_{b \in B} h(a,b)g(b,c) + \sum_{b' \in B} h(a',b')g(b',c') \right]$$
$$= h_A(a,a')[(g \circ h)(a,c) + (g \circ h)(a',c')],$$

as desired. $\square$

Another possibilty is to add weights. Let $w \in R^A$. A symmetric and reflexive R-valued relation $h \in R^{A \times A}$ is an *R-valued equivalence relation with weight w* if and only if it satisfies $h(a,a')w(a')h(a',a'') \preceq h(a,a'')$ for all $a, a', a'' \in A$. For the case of $R = (\mathbb{R}^+ \cup \{\infty\}, \wedge, +)$, this construction appears in [403] and in [127].

Quasimetrics and pseudometrics. We noted above that R-valued equivalence relations can be considered as the duals of extended pseudometrics. It is worth noting, in passing, that semiring-valued pseudometrics defined on semimodules are of interest in themselves.

Let R be a partially-ordered semiring and let M be a nonempty set. A function $\rho \in R^{M \times M}$ is a *quasimetric* on M with values in R if and only if the following conditions are satisfied:

(1) $\rho(m,m) = 0$ for all $m \in M$;
(2) $\rho(m,m'') \leq \rho(m,m') + \rho(m',m'')$ for all $m, m', m'' \in M$.

If the additional condition

(3) $\rho(m,m') = \rho(m',m)$ for all $m, m' \in M$

is satisfied, then ρ is a *pseudometric* on M. If $(R, \vee, \wedge)$ is a bounded distributive lattice, if $h \in R^{M \times M}$ is an R-valued equivalence relation, and if $\delta \colon R \to R$ is a complementation then δh is an $(R, \wedge, \vee)$-valued quasimetric on M.

All topological spaces can be defined using quasimetrics and pseudometrics with values in suitable semirings [214].

If the semiring R is a complete lattice-ordered semiring and if $\theta \colon (R, +) \to (R, \cdot)$ is a function satisfying

(i) $\theta(0) = 1$; and
(ii) $\theta(r)\theta(r') \leq \theta(r + r')$ for all $r, r' \in R$

then, for any left R-semimodule M, we can define a quasimetric ρ_θ on M with values in R by setting

$$\rho_\theta(m, m') = \bigwedge \{r \in R \mid r \leq \theta(r)m'\}$$

for all $m, m' \in M$. Indeed, since lattice-ordered semirings are positive [146, Proposition 19.13] we immediately see that $\rho_\theta(m, m) = 0$ for all $m \in M$. Moreover, if $m, m', m'' \in M$ and if $\rho_\theta(m, m') = r_1$ while $\rho_\theta(m', m'') = r_2$ then

$$m \leq \theta(r_1)m' \leq \theta(r_1)\theta(r_2)m'' \leq \theta(r_1 + r_2)m''$$

so $\rho_\theta(m, m'') \leq \rho_\theta(m, m') + \rho_\theta(m', m'')$. Moreover, we can also define a pseudometric δ_θ on M with values in R by setting $\delta_\theta(m, m') = \rho_\theta(m, m') + \rho_\theta(m', m)$. Note that if $\alpha \colon M_1 \to M_2$ is a homomorphism of left M-semimodules and if $m, m' \in M_1$ then $\rho_\theta(m\alpha, m'\alpha) \leq \rho_\theta(m, m')$. Moreover, we have equality if α is monic.

Let R be a complete lattice-ordered semiring and let M be a nonempty set. Given a quasimetric $\rho \in R^{M \times M}$ and a nonempty subset E of R closed under taking meets, we define

$$W_{\rho, e}(m) = \{m' \in M \mid \rho(m, m') \leq e\}$$

for all $e \in E$ and all $m \in M$. The family of all subsets of M of this form is closed under taking finite intersections and so forms a basis for a topology on M. Compare this construction with that given in [144].

(I) EXAMPLE. [150] Let R be a ring and let $ideal(R)$ the complete lattice-ordered semiring of all (two-sided) ideals of R. Let M be a left R-module and, as above, let $sub(M)$ denote the set of all submodules of M. Then $(sub(M), +)$ is a left $ideal(R)$-semimodule, with the product of an ideal of R and a submodule of M being defined in the standard manner. If $H, I \in ideal(R)$, set $(H : I) = \{r \in R \mid rI \subseteq H\}$. Then $(H : (0)) = R$ for all $H \in ideal(R)$ and

$$(H : I)(H : I') \subseteq (H : I) \cap (H : I') \subseteq (H : I + I')$$

for all $H, I, I' \in ideal(R)$. This means that each ideal H of R defines a function $\theta_H: I \mapsto (H:I)$ from $ideal(R)$ to itself which satisfies conditions (i) and (ii) above and so defines a $ideal(R)$-valued quasimetric ρ_H on $sub(M)$ given as follows:

$$\rho_H(N, N') = \bigcap \{I \in ideal(R) \mid N \subseteq (H:I)N'\}.$$

Note that these quasimetrics are *compatible* in the following sense: if M_1 is a submodule of M_2 then $sub(M_1) \subseteq sub(M_2)$ and ρ_H on $sub(M_1)$ is merely the restriction of ρ_H as defined on $sub(M_2)$.

(II) EXAMPLE. [150] Let R be a ring and let $R - fil$ the complete lattice-ordered semiring of all topologizing filters of left ideals of R, in which the induced order is reverse inclusion. If $\kappa_1, \kappa_2 \in R - fil$ then the right residual $\kappa_1^{-1}\kappa_2$ is the element of $R - fil$ defined by

$$\kappa_1^{-1}\kappa_2 = \bigcap \{\kappa \in R - fil \mid \kappa_1 \kappa \supseteq \kappa_2\}.$$

By Proposition 3.6 and Proposition 4.14 of [142], we see that

$$(\kappa_1 \cap \kappa_2)^{-1}\kappa = \kappa_1^{-1}\kappa \vee \kappa_2^{-1}\kappa \subseteq (\kappa_1^{-1}\kappa)(\kappa_2^{-1}\kappa)$$

for all $\kappa_1, \kappa_2, \kappa \in R - fil$. Moreover, the $\cap$-neutral element of $R - fil$ is the filter $\eta[0]$ of all left ideals of R and for any $\kappa \in R - fil$ we have $\eta[0]^{-1}\kappa = \{R\}$, which is just the neutral element of $R - fil$ with respect to multiplication (i.e. the Gabriel product). Thus we see that each $\kappa \in R - fil$ defines a function $\theta_\kappa: R - fil \to R - fil$ satisfying conditions (i) and (ii) above.

Let M be a left R-module and let $sub(M)$ again denote the lattice of all submodules of M. Following Example 13.13 of [146], we note that $(sub(M), \cap)$ can be considered as a right semimodule over $R - fil$ where, for each $N \in sub(M)$ and each $\kappa \in R - fil$, we let $N\kappa$ be the κ-purification of N in M. That is to say, an element $m \in M$ belongs to $N\kappa$ if and only if there exists a left ideal I of R belonging to κ and satisfying $Im \subseteq N$. As above, we thus have a quasimetric ρ_κ defined as follows: if N and N' are submodules of M then

$$\rho_\kappa(N, N') = \bigvee \{\kappa_1 \in R - fil \mid N \supseteq N'(\kappa_1^{-1}\kappa)\}$$

Again, these quasimetrics are compatible.

Functions between semiring powers. Let R be a semiring. Any function $u: A \to B$ between nonempty sets defines an R-valued function h_u between A and B by setting

$$h_u(a, b) = \begin{cases} 1 & \text{if } u(a) = b \\ 0 & \text{otherwise.} \end{cases}$$

Therefore, if $f \in R^A$ we have

$$h_u[f]: b \mapsto \sum_{a \in A} f(a)h_u(a,b) = \sum_{u(a)=b} f(a)$$

while if $g \in R^B$ we have

$$h_u^{-1}[g]: a \mapsto \sum_{b \in B} h_u(a,b)g(b) = gu(a).$$

Thus, $u \in Mor(f,g)$ if and only if $f \preceq h_u^{-1}[g]$ in R^A.

We note, in particular, that if the function $u: A \to B$ is bijective then

$$h_u[f]: b \mapsto (fu^{-1})(b)$$

for all $b \in B$.

Thus, if R is additively idempotent and if $u: A \to B$ is a function then $h_u(\chi_{A'}) = \chi_{u(A')}$ for all $A' \subseteq A$ and $h_u^{-1}(\chi_{B'}) = \chi_{u^{-1}(B')}$ for all $B' \subseteq B$.

(2.8) Proposition. *Let R be a complete semiring and let $u: A \to B$ be a function between nonempty sets. Then:*
(1) $h_u[h_u^{-1}[g]] \preceq g$ *for all $g \in R^B$; and*
(2) $f \preceq h_u^{-1}[h_u[f]]$ *for all $f \in R^A$.*

PROOF. (1) If $b_0 \in B$

$$h_u[h_u^{-1}[g]]: b_0 \mapsto \sum_{a \in A} h_u^{-1}[g](a)h_u(a,b_0) = \sum_{a \in A}\sum_{b \in B} h_u(a,b)g(b)h_u(a,b_0).$$

But this sum equals 0_R except in the case $b = b_0 = u(a)$, in which case it equals $g(b_0)$. Thus, if $g_1 \in R^B$ is the function defined by

$$g_1(b) = \begin{cases} 0 & \text{if } b \in im(u) \\ g(b) & \text{otherwise.} \end{cases}$$

then $h_u[h_u^{-1}[g]] + g_1 = g$ and so $h_u[h_u^{-1}[g]] \preceq g$.
(2) If $a_0 \in A$ then

$$h_u^{-1}[h_u[f]]: a_0 \mapsto \sum_{b \in B} h_u(a_0,b)h_u[f](b)$$
$$= \sum_{b \in B}\sum_{a \in A} h_u(a_0,b)f(a)h_u(a,b)$$
$$= \sum_{f(a)=f(a_0)} f(a)$$

and so $f \preceq h_u^{-1}[h_u[f]]$. □

Note that Proposition 2.8 implies that if R is a complete semiring and if $u\colon A \to B$ is a function between nonempty sets then the function $g \mapsto h_u[h_u^{-1}[g]]$ is an interior operator on R^B and the function $f \mapsto h_u^{-1}[h_u[f]]$ is a closure operator on R^A.

Let R be a semiring and let $u\colon A \to B$ be a function between nonempty sets. Then $f \in R^A$ is *u-stable* if and only if $f(a_1) = f(a_2)$ whenever $u(a_1) = u(a_2)$.

(2.9) PROPOSITION. *Let R be an additively-idempotent semiring and let A and B be finite sets. If $u\colon A \to B$ is a surjective map then there exists a bijective correspondence between R^B and the set of all u-stable elements of R^A.*

PROOF. If $g \in R^B$ then $h_u^{-1}[g] \in R^A$ is easily seen to be u-stable and

$$h_u[h_u^{-1}[g]] = g$$

since, for all $b \in B$, we have $h_u[h_u^{-1}[g]](b) = \sum_{u(a)=b} gu(a) = g(b)$ by the u-stability of g and the additive idempotence of R.

Now suppose that $f_1, f_2 \in R^A$ are u-stable functions satisfying $h_u(f_1) = h_u(f_2)$. If $a_0 \in A$ and $b_0 = u(a_0) \in B$ then, by the additive idempotence of R, we have

$$f_1(a_0) = \sum_{u(a)=b_0} f_1(a) = h_u(f_1)(b_0) = h_u(f_2)(b_0) = f_2(a_0)$$

and so $f_1 = f_2$. □

Other constructions. Let R be a semiring partially-ordered with respect to $\leq$ and let A be a nonempty set. A function $f \mapsto f^c$ from R^A to itself is a *closure operator* with respect to $\leq$ if and only if the following conditions are satisfied:

(1) $f \leq f^c = f^{cc}$ for all $f \in R^A$;
(2) If $f \leq g$ in R^A then $f^c \leq g^c$.

In particular, we will be interested in closure operators with respect to $\preceq$ on difference-ordered semirings. A closure operator is *linear* if it satisfies the additional condition

(3) $(f + g)^c = f^c + g^c$ for all $f, g \in R^A$.

Closure operators allow for the construction of generalized topological structures on A [128] and serve as interesting models for databases [53, 65].

As an example, let R be a complete semiring and let A be a nonempty set. Any nonempty subset U of R^A defines a function $c_U\colon R^A \to R^A$ which assigns to each $g \in R^A$ the function $c_U[g] \in R^A$ defined by $c_U[g] = \wedge \{f \in U \mid f \succeq g\}$. (By the usual convention, the meet of any empty subset of R^A is taken to be the unique maximal element of R^A.) We immediately note that this is a linear closure operator on R^A. This construction is the basis for Pavelka's fuzzy logic [124, 313], in which the function c_U acts as a consequence operator.

If $f \mapsto f^c$ is a closure operator on R^A then an element f of R^A satisfying $f = f^c$ is *c-closed*. An element f of R^A is *c-free* if and only if $a \notin supp((f\neg p_{a,1})^c)$ for all $a \in supp(f)$. If $f \leq g$ in R^A then f is *maximally c-free in g* if and only if it is c-fee and there is no c-free $f_1 \in R^A$ satisfying $f < f_1 \leq g$. Also, f is a *c-basis* for g if and only if f is c-free and $f^c = g$.

Let A be a nonempty set and let $\mathcal{A}$ be a nonempty family of nonempty subsets of A satisfying the condition that if $B, B' \in \mathcal{A}$ then $B \cap B' \in \mathcal{A}$. If R is a semiring then $R^B \cap R^{B'} = \varnothing$ for all $B \neq B'$ in $\mathcal{A}$. Set $R^{\mathcal{A}} = \bigcup \{R^B \mid B \in \mathcal{A}\}$. Thus, for each $f \in R^{\mathcal{A}}$ there exists a unique $B \in \mathcal{A}$ such that $f \in R^B$. This subset B of A is called the *domain* of f and will be denoted by $dom(f)$. We now define operations of addition and multiplication on $R^{\mathcal{A}}$ as follows:

(1) If $f, g \in R^{\mathcal{A}}$ then $dom(f + g) = dom(f) \cap dom(g)$ and $(f + g)(a) = f(a) + g(a)$ for all $a \in dom(f + g)$.
(2) If $f, g \in R^{\mathcal{A}}$ then $dom(fg) = dom(f) \cap dom(g)$ and $(fg)(a) = f(a)g(a)$ for all $a \in dom(fg)$.

(2.10) PROPOSITION. *Let A be a nonempty set and let $\mathcal{A}$ be a nonempty family of nonempty subsets of A satisfying the condition that if $B, B' \in \mathcal{A}$ then $B \cap B' \in \mathcal{A}$. If R is a semiring then $R^{\mathcal{A}}$ is also a semiring.*

PROOF. It is straightforward to check all of the conditions in the definition of a semiring. Notice that the additive identity of $R^{\mathcal{A}}$ is the function $a \mapsto 0$ having domain A and the multiplicative identity in $R^{\mathcal{A}}$ is the function $a \mapsto 1$ having domain A. □

Thus, in particular, we see that if A is a nonempty set then $R^A = R^{\{A\}}$.

For example, let $\mathcal{A}$ be the set of all cofinite subsets of $\mathbb{R}$. The semiring $\mathbb{R}^{\mathcal{A}}$ is fundamental in the study of elementary calculus, as was pointed out in [117].

Let R and S be semirings. An *R-gradation* on S is a function $\varphi \colon R \to S$ satisfying the condition that $r \preceq r'$ implies $\varphi(r) \succeq \varphi(r')$. In particular, if A is a nonempty set, then an *R-gradation* on A is an R-gradation on the semiring $\mathbb{P}(A)$ of all subsets of A. A pair (A, φ) consisting of a nonempty set A together with an R-gradation on A is an *R-graded set*. See [182] for a study of $\mathbb{I}$-graded sets.

If R and S are CLO-semirings then an R-gradation φ on S is *complete* if and only if

$$\varphi\left(\sum_{i \in \Omega} r_i\right) = \bigwedge_{i \in \Omega} \varphi(r_i)$$

for every family $\{r_i \mid i \in \Omega\}$ of elements of R.

If R is a semiring, if $\{S_i \mid i \in \Omega\}$ is a family of semirings and if we are give an R-gradation $\varphi_i \colon R \to S_i$ on S_i for each $i \in \Omega$, then the function $\varphi \colon R \to S = \times_{i \in \Omega} S_i$ defined by $\varphi(r)(i) = \varphi_i(r)$ for all $i \in \Omega$ is an R-gradation on S.

Also, let R and S be semirings and let $\varphi \colon R \to S$ and $\varphi' \colon R \to S$ be R-gradations of S. Then $(\varphi + \varphi') \colon R \to S$ is an R-gradiation of S. More generally, let R be a semiring and let S be a complete semiring. Let Ω be a nonempty set and for each $i \in \Omega$ let $\varphi_i \colon R \to S$ be an R-graduation on S. Then the function $\varphi \colon R \to S$ defined by $\varphi \colon r \mapsto \sum_{i \in \Omega} \varphi_i(r)$ is an R-gradiation of S.

Let R and S be simple semirings and let $\delta \colon R \to R$ and $\zeta \colon S \to S$ be complementations. If $\varphi \colon R \to S$ is an R-gradation on S then so is $\zeta \varphi \delta \colon R \to S$. In particular, if (A, φ) is an R-graded set, where R is a simple semiring on which we are given a complementation δ, then (A, ψ) is an R-graded set, where ψ is defined by $\psi \colon r \mapsto A \setminus \varphi \delta(r)$ for each $r \in R$.

The most important example of a gradation which we will need is the following: Let R be a semiring and let A be a nonempty set. For $f \in R^A$ and $r \in R$, set

$$L_r(f) = \{a \in A \mid f(a) \succeq r\}.$$

Thus, for example, if $R = (\mathbb{R} \cup \{\infty\}, \wedge, +)$ and $f \in R^A$ then $L_r(f)$ is the f-preimage of the closed ball around 0 of radius r.

We note immediately that if $r \succeq r'$ in R then $L_r(f) \subseteq L_{r'}(f)$ and so the function $r \mapsto L_r(f)$ is an R-gradation on A for each $f \in R^A$. Moreover, $L_0(f) = A$ and if R has an infinite element ∞ then $L_\infty(f) \subseteq L_r(f)$ for all $r \in R$. Also, $L_{r+r'}(f) \subseteq L_r(f) \cap L_{r'}(f)$ for all $r, r' \in R$. If R is additively idempotent then we in fact have equality. Indeed, in this case we note that if $a \in L_r(f) \cap L_{r'}(f)$ then there exist $s, s' \in R$ such that $f(a) = r + s = r' + s'$ and so $f(a) = f(a) + f(a) = r + r' + s + s'$, proving that $a \in L_{r+r'}(f)$.

Thus, in particular, choose $R = (\mathbb{R}^+ \cup \{\infty\}, \wedge, +)$ and let $d \in R^{A \times A}$ be an extended pseudometric on a nonempty set A. That is to say, d is an R-valued equivalence relation on A. For a fixed point $a_0 \in A$, we then have a function $f \in R^A$ defined by $f \colon a \mapsto d(a_0, a)$. If $r \in R$ then $L_r(f) = \{a \in A \mid d(a_0, a) \succeq r\}$ which, when stated in terms of the usual order $\leq$ on $\mathbb{R}$, is just $\{a \in A \mid d(a_0, a) \leq r\}$. In other words, $L_r(f)$ is just the closed ball of radius r around the point a_0.

Given a nonempty set A, Swamy and Raju [374] consider elements f of $\mathbb{I}^A$ in which the sets $L_r(f)$, for each $r \in \mathbb{I}$, must lie in a designated family $\mathcal{S}$ of subsets of A.

To take another example, let A be a nonempty set and let R be a CLO-semiring. Every R-valued subset $f \in R^A$ of A defines a complete $\mathbb{P}(A)$-gradation w_f of R

by setting
$$w_f : B \mapsto \begin{cases} \bigvee_{a \in A} f(a) & \text{if } B = \emptyset \\ \bigwedge_{b \in B} f(b) & \text{otherwise} \end{cases}.$$

Note that if $\{\varphi_i \mid i \in \Omega\}$ is an arbitrary nonempty family of R-gradations on A then the function
$$\bigcup_{i \in \Omega} \varphi_i : r \mapsto \bigcup_{i \in \Omega} \varphi_i(r).$$
and the function
$$\bigcap_{i \in \Omega} \varphi_i : r \mapsto \bigcap_{i \in \Omega} \varphi_i(r).$$
are also R-gradations on A.

If A is a nonempty set then any $f \in \mathbb{I}^A$ defines two $\mathbb{I}$-gradations on A, namely
$$\overline{\varphi_f} : r \mapsto \{a \in A \mid f(a) \geq r\}$$
and
$$\varphi_f : r \mapsto \{a \in R \mid f(a) > r\}.$$

Herencia [182] characterizes those $\mathbb{I}$-gradations on A which have this form, as follows: If ψ is an $\mathbb{I}$-gradation on A then

(1) $\psi = \overline{\varphi_f}$ for some $f \in \mathbb{I}^A$ if and only if $\psi(0) = A$ and, for all $0 < r \leq 1$ in $\mathbb{I}$, we have $\psi(r) = \cap\{\psi(r') \mid 0 \leq r' < r\}$.
(2) $\psi = \varphi_f$ for some $f \in \mathbb{I}^A$ if and only if $\psi(1) = \emptyset$ and, for all $0 \leq r < 1$ in $\mathbb{I}$, we have $\psi(r) = \cup\{\psi(r') \mid r < r' \leq 1\}$.

The following is a slight generalization of a result given in [298].

(2.11) PROPOSITION. *Let R be a complete semiring every element of which is compact, let A be a nonempty set, and let φ be a complete R-gradation on A. Then there exists an $f \in R^A$ such that $\varphi(r) = L_r(f)$ for all $r \in R$.*

PROOF. Let $f \in R^A$ be the function defined by $f : a \mapsto \sum\{t \in R \mid a \in \varphi(t)\}$. If $r \in R$ and $a \in \varphi(r)$ then $f(a) \succeq r$ and so $a \in L_r(f)$. Thus $\varphi(r) \subseteq L_r(f)$ for all $r \in R$.

Conversely, assume that $a \in L_r(f)$. Then $f(a) \succeq r$. If $f(a) = r$ then $\varphi(r) = \cap\{\varphi(t) \mid a \in \varphi(t)\}$ and so $a \in \varphi(r)$. If $f(a) \succ r$ then, by compactness, thee exist elements $t_1, \ldots, t_n$ of R satisfying $a \in \varphi(t_i)$ for all $1 \leq i \leq n$ and $r \preceq t_1 + \cdots + t_n$. Then $a \in \varphi(t_1) \cap \ldots \cap \varphi(t_n) = \varphi(t_1 + \cdots + t_n) \subseteq \varphi(r)$. Hence we have equality. $\square$

3. Change of Base Semirings

If R and S are semirings then a function $\gamma: R \to S$ is a *morphism of semirings* if and only if the following conditions are satisfied:

(1) $\gamma(r + r') = \gamma(r) + \gamma(r')$ and $\gamma(rr') = \gamma(r)\gamma(r')$ for all $r, r' \in R$;
(2) $\gamma(0_R) = 0_S$;
(3) $\gamma(1_R) = 1_S$.

Note that if $\gamma: R \to S$ is a morphism of semirings and $r \ll r'$ in R then $\gamma(r) \ll \gamma(r')$ in S and that if $r \preceq r'$ in R then $\gamma(r) \preceq \gamma(r')$ in S.

If the semirings R and S are complete and if a morphism of semirings $\gamma: R \to S$ satisfies the additional condition

(4) $\gamma(\sum_{i \in \Omega} r_i) = \sum_{i \in \Omega} \gamma(r_i)$ for all nonempty index sets Ω

then γ is a *morphism of complete semirings*.

Let us look at some examples.

(I) EXAMPLE. Let R be a semiring and let A be a nonempty set. A function $f \in R^A$ is *proper* if and only if $im(f)$ is not a singleton. Otherwise, f is *improper*. Clearly the set of all improper elements of R^A is a subsemiring of R^A. For each $r \in R$ let $f_r \in R^A$ be the constant function $f_r: a \mapsto r$. Then the map $r \mapsto f_r$ is a monic morphism of semirings from R to R^A, the image of which is the subsemiring of all improper elements of R^A.

(II) EXAMPLE. If R is a commutative multiplicatively-cancellative semiring then for each positive integer n the function $\gamma_n: R \to R$ given by $\gamma_n: a \mapsto a^n$ is a morphism of semirings.

(III) EXAMPLE. [394] A morphism from a semiring R to $\mathbb{B}$ is called a *character* of R. A commutative semiring is a ring precisely when it has no characters. Thus, a commutative zerosumfree semiring certainly does have characters. The commutativity condition here is necessary. Indeed, let R be the noncommutative semiring $\mathcal{M}_2(\mathbb{B})$ and let us assume that $\gamma \in char(R)$. We claim that

$$\gamma\left(\begin{bmatrix} 1 & 0 \\ 0 & 0 \end{bmatrix}\right) = 0.$$

Indeed, if

$$\gamma\left(\begin{bmatrix} 1 & 0 \\ 0 & 0 \end{bmatrix}\right) = 1$$

then

$$\gamma\left(\begin{bmatrix} 0 & 0 \\ 1 & 0 \end{bmatrix}\right) = 0$$

since $\begin{bmatrix} 1 & 0 \\ 0 & 0 \end{bmatrix} \begin{bmatrix} 0 & 1 \\ 0 & 0 \end{bmatrix} = \begin{bmatrix} 0 & 0 \\ 0 & 0 \end{bmatrix}$. Thus

$$\gamma\left(\begin{bmatrix} 1 & 0 \\ 1 & 0 \end{bmatrix}\right) = \gamma\left(\begin{bmatrix} 1 & 0 \\ 0 & 0 \end{bmatrix} + \begin{bmatrix} 0 & 0 \\ 1 & 0 \end{bmatrix}\right) = 1 + 0 = 1.$$

But $\begin{bmatrix} 1 & 0 \\ 1 & 0 \end{bmatrix} = \begin{bmatrix} 1 & 1 \\ 0 & 1 \end{bmatrix} \begin{bmatrix} 0 & 0 \\ 1 & 0 \end{bmatrix}$ and so

$$\gamma\left(\begin{bmatrix} 1 & 0 \\ 1 & 0 \end{bmatrix}\right) = \gamma\left(\begin{bmatrix} 1 & 1 \\ 0 & 1 \end{bmatrix}\right) \cdot \gamma\left(\begin{bmatrix} 0 & 0 \\ 1 & 0 \end{bmatrix}\right) = 0,$$

which is a contradiction that establishes the claim. Similarly, we must have

$$\gamma\left(\begin{bmatrix} 0 & 0 \\ 0 & 1 \end{bmatrix}\right) = 0.$$

Therefore

$$\gamma\left(\begin{bmatrix} 1 & 0 \\ 0 & 1 \end{bmatrix}\right) = \gamma\left(\begin{bmatrix} 1 & 0 \\ 0 & 0 \end{bmatrix}\right) + \gamma\left(\begin{bmatrix} 0 & 0 \\ 0 & 1 \end{bmatrix}\right) = 0,$$

which is also a contradiction. See [153] for further details.

(IV) EXAMPLE. Let R and S be semirings and let A be a nonempty set. If D is a nonempty set and if we have a morphism of semirings $\gamma: R^A \to (S^A)^D$ then each $d \in D$ defines a morphism of semirings $\gamma_d: R^A \to S^A$, which is just the composition of γ and the *evaluation function* $\epsilon_d: T \to T(d)$ from $(S^A)^D$ to S^A.

(V) EXAMPLE. Let R be a commutative ring and let $(ideal(R), +, \cdot)$ be the semiring of all ideals of R. Let $Zar(R)$ be the family of all subsets of $spec(R)$ closed in the Zariski topology. Then $(Zar(R), \cap, \cup)$ is a semiring and the map $ideal(R) \to Zar(R)$ defined by $I \mapsto V(I)$ is a surjective morphism of semirings.

(VI) EXAMPLE. [120] Let $R = \{0\} \cup \{(a,b) \mid 0 < a, b \in \mathbb{R}\}$ on which we define addition given by:
(1) $0 + r = r = r + 0$ for all $r \in R$.
(2) If $(a,b), (a',b') \in R \setminus \{0\}$ then

$$(a,b) + (a',b') = \begin{cases} (a+a', b) & \text{if } b = b' \\ (a,b) & \text{if } b > b' \\ (a',b') & \text{if } b < b' \end{cases}.$$

and multiplication given by $(a,b) \cdot (a',b') = (aa', bb')$. Then R is a zerosumfree semifield and we have an injective morphism of semirings $\gamma \colon (\mathbb{R}^+, +, \cdot) \to R$ given by

$$\gamma \colon a \mapsto \begin{cases} (a, 1) & \text{if } a \neq 0 \\ 0 & \text{otherwise} \end{cases}.$$

We also have a surjective morphism of semirings $\gamma' \colon R \to (\mathbb{R}^+, \vee, \cdot)$ given by

$$\gamma' \colon (a,b) \mapsto \begin{cases} b & \text{if } (a,b) \neq 0 \\ 0 & \text{otherwise} \end{cases}.$$

See [106, 120] for applications.

If A and B are nonempty sets and if $\theta \colon A \to B$ is a function then any morphism of semirings $\gamma \colon R \to S$ defines a morphism of semirings $\gamma^\theta \colon R^B \to S^A$ by $(\gamma^\theta f)(a) = \gamma(f(\theta(a)))$. In particular, if $A \subseteq B$ are nonempty sets and if R is a semiring then we have a canonical morphism of semirings $R^B \to R^A$ given by restriction of functions. Moreover, any morphism of semirings $\gamma \colon R \to S$ induces a morphism of semirings $\gamma^A \colon R^A \to S^A$ defined by $\gamma(f) \colon a \mapsto \gamma(f(a))$ for all $f \in R^A$. In particular, if $R_{(0)}$ is the basic semiring of a semiring R then, for any nonempty set A, we see that $R_{(0)}^A$ is a subsemiring of R^A. The elements of this subsemiring are called the *crisp* elements of R^A. Thus, for example, if R is additively idempotent then the crisp elements of R^A are precisely the functions in $\mathbb{B}^A$.

Now let R and S be semirings and let A be a nonempty set. If h is an S-valued relation on R then we have a canonical function $\rho_{h,A} \colon R^A \to (S^A)^R$ defined by

$$\rho_{h,A}(f)(r) \colon a \mapsto h(f(a), r).$$

Thus, using example (IV) above, we see that if R and S are semirings and if $h \in S^{A \times A}$ is a strongly-symmetric relation then each $r \in R$ defines a *projection function* $\pi_r \colon R^A \to S^A$ which is just $\rho_{h,A}$ composed with the *evaluation function* $\varepsilon_r \colon T \mapsto T(r)$ from $(S^A)^R$ to S^A. Moreover, we also have the following:

(3.1) PROPOSITION. *Let R and S be semirings and let A be a nonempty set. If h is a strongly symmetric S-valued relation on R then the function $\rho_{h,A} \colon R^A \to (S^A)^R$ is monic.*

PROOF. Suppose that $f, g \in R^A$ satisfy $\rho_{h,A}(f) = \rho_{h,A}(g)$. Then for each $a \in A$ we have $h(f(a), g(a)) = \rho_{h,A}(f)(g(a))(a) = \rho_{h,A}(g)(g(a))(a) = h(g(a), g(a)) = 0$, which implies that $f(a) = g(a)$. Hence $f = g$. □

We are interested, in particular, in knowing when such functions are semiring homomorphisms.

We now consider a construction based on some observations of [172]. A subset G of a complete semiring R is an *additive set of generators* if and only if for each $0 \neq r \in R$ there exists a family $\{r_i \mid i \in \Omega\}$ of elements of G such that $r = \sum_{i \in \Omega} r_i$. Of course, R is always an additive set of generators for itself.

Consider a compact element u of a CLO-semiring R. (For example, if S is a commutative semiring and $a \in S$ then Sa is a compact element of the semiring $R = ideal(S)$ of all ideals of S.) Then the semiring R is *algebraic* if and only if the set of all compact elements of R is an additive set of generators.

Let R be a complete semiring with necessary summation and let G be a nonempty subset of R. Following [172], we note that there is a canonical function $\rho_A \colon R^A \to (\mathbb{B}^A)^G$ defined by

$$\rho_A(f)(r) \colon a \mapsto \begin{cases} 0 & \text{if } f(a) \leq r \\ \infty & \text{otherwise} \end{cases}$$

for all $r \in G$.

A similar function is the function $\rho'_A \colon R^A \to (\mathbb{B}^A)^G$ defined by

$$\rho'_A(f)(r) \colon a \mapsto \begin{cases} 0 & \text{if } f(a) = r \\ \infty & \text{otherwise} \end{cases}$$

for all $r \in G$.

(3.2) PROPOSITION. *If R is a semiring with necessary summation having an additive set of generators G and if A is a nonempty set then the function $\rho_A \colon R^A \to (\mathbb{B}^A)^G$ is monic.*

PROOF. Suppose that $\rho_A(f) = \rho_A(g)$ for $f, g \in R^A$. If $a \in A$ then we write $f(a) = \sum_{i \in \Omega} u_i$ and $g(a) = \sum_{j \in \Lambda} v_j$, where the u_i and v_j belong to G. For each $j \in \Lambda$ we then have $\rho_A(f)(v_j)(a) = \rho_A(g)(v_j)(a) = 0$ since, by the difference order on R, $v_j \leq g(a)$. Therefore $v_j \leq f(a)$ for all $j \in \Omega$ and so $g(a) = \sum_{j \in \Lambda} v_j \leq f(a)$. A similar argument shows that $f(a) \leq g(a)$ and so we have equality for each $a \in A$, proving that $f = g$. □

(3.3) PROPOSITION. *If R is a complete semiring with necessary summation having an additive set of generators G and if A is a nonempty set then the function $\rho_A \colon R^A \to (\mathbb{B}^A)^G$ commutes with arbitrary sums (= joins) and finite meets.*

PROOF. Let $\{f_i \mid i \in \Omega\}$ be a family of elements of R^A, let $r \in G$, and let $a \in A$. Then

$$\left[\rho_A\left(\sum_{i \in \Omega} f_i\right)\right](r)(a) = 0 \Leftrightarrow \sum_{i \in \Omega} f_i(a) \leq r$$

$$\Leftrightarrow \bigvee_{i \in \Omega} f_i(a) \leq r$$

$$\Leftrightarrow f_i(a) \leq r \text{ for all } i \in \Omega$$

$$\Leftrightarrow \left[\bigvee_{i \in \Omega} \rho_A(f_i)\right](r)(a) = 0$$

$$\Leftrightarrow \left[\sum_{i \in \Omega} \rho_A(f_i)\right](r)(a) = 0$$

Thus $\rho_A(\sum_{i \in \Omega} f_i) = \sum_{i \in \Omega} \rho_A(f_i)$. Similarly, if $f, g \in R^A$ then $\rho_A(f \wedge g) = \rho_A(f) \wedge \rho_A(g)$. □

Commuting with products is more of a problem. Recall that an element r of a partially-ordered semiring R is *prime* if and only if $r_1 r_2 \leq r$ implies that $r_1 \leq r$ or $r_2 \leq r$. For example, if $R = (\mathbb{I}, \vee, \wedge)$ then every element of R is prime since R is totally ordered.

(3.4) PROPOSITION. *Let R be a complete semiring with necessary summation and suppose that G is an additive set of generators for R each member of which is prime. Then the function $\rho_A \colon R^A \to (\mathbb{B}^A)^G$ commutes with finite products.*

PROOF. Let $f, g \in R^A$, $r \in G$, and $a \in A$. Then

$$\rho_A(fg)(r)(a) = 0 \Leftrightarrow (fg)(a) \leq r$$

$$\Leftrightarrow f(a)g(a) \leq r$$

$$\Leftrightarrow f(a) \leq r \text{ or } g(a) \leq r$$

$$\Leftrightarrow \rho_A(f)(r)(a) = 0 \text{ or } \rho_A(g)(r)(a) = 0$$

$$\Leftrightarrow [\rho_A(f)\rho_A(g)](r)(a) = 0$$

and so $\rho_A(fg) = \rho_A(f)\rho_A(g)$. □

If $\gamma \colon R \to S$ is a morphism of complete semirings we define the functions $\gamma^{(-1)}$ and $\gamma^{[-1]}$ from S to R by

$$\gamma^{(-1)} \colon s \mapsto \sum \{r \in R \mid \gamma(r) \ll s\}$$

and
$$\gamma^{[-1]}: s \mapsto \sum\{r \in R \mid \gamma(r) \preceq s\}.$$

Note that $r \preceq \gamma^{[-1]}\gamma(r)$ for each $r \in R$.

4. Convolutions

We now begin to consider powers of semirings with a certain structure on the exponent set by looking at semirings of the form R^M, where R is a semiring and $(M, *)$ is a semigroup. Such structures show up in various contexts, and later we shall see many specific examples.

A semigroup $(M, *)$ is *finitary* if and only if each element $m \in M$ can be written in the form $m' * m''$ in only finitely-many ways. If $(M, *)$ is a finitary semigroup then, for any semiring R, the set R^M is equipped with a naturally-defined operation $\langle * \rangle$, called *$*$-convolution*, and defined by

(*) $$f\langle * \rangle g \colon m \mapsto \sum_{m'*m''=m} f(m')g(m'')$$

for all $m \in M$, subject to the usual convention that the sum of the empty set of elements of R is taken to be 0_R. The study of such operations harks back to Dirichlet, who considered $\mathbb{C}^M$, where M is the multiplicative monoid of positive integers. His work opened the door to an extensive theory of convolutions of what came to be known as arithmetic functions.

If U and V are nonempty subsets of R^M then, following the usual convention, we set $U\langle * \rangle V = \{ f\langle * \rangle g \mid f \in U, g \in V \}$.

It is easy to verify that $(R^M, +, \langle * \rangle)$ is a hemiring, called the *power algebra* over R defined by M. It is not necessarily a semiring; in order for it to be so there must exist a function $e \in R^M$ such that, for all elements $m \neq n$ of M, we have

$$\sum_{m'*m=m} e(m') = 1 = \sum_{m*m''=m} e(m'')$$

and

$$\sum_{m'*m=n} e(m') = 0 = \sum_{m*m''=n} e(m'').$$

See Proposition 4.10 of [146]. For example, if M is a monoid having identity element m_0 and if the semiring R is additively-idempotent, we can define e to be the characteristic function on $\{m_0\}$.

The problem of finding necessary and sufficient conditions for R^M to have a multiplicative identity is difficult, even if R is taken to be a ring, and has been well-studied in the literature. See [322] for details and a list of references; also refer to [344].

If the semiring R is complete then we can dispense with the requirement that M be finitary. If either the semiring R is complete or if M is finitary, we will say that the pair (R, M) forms a *convolution context*. In most of what follows, we will assume that we are working within such a context. Note that if (R, M) is a convolution context, then so is (R, N) for any subsemigroup N of M.

Of course, if we do not know that we are in a convolution context, we can always restrict ourselves to the subhemiring $R^{(M)}$ of R^M consisting of all elements of R^M having finite support. This is the *restricted convolution algebra* over R defined by M. If R^M has a multiplicative identity, we would want to adjoin it to $R^{(M)}$ as well, in order to get a subsemiring. A further generalization, which we will not as a rule have to resort to, replaces the semigroup M by a partial semigroup, i.e. a nonempty set on we have an operation $*$ only partially-defined.

Clearly, in the above situations, the operation $\langle * \rangle$ on R^M is commutative if $(M, *)$ is a commutative semigroup and R is a commutative semiring.

Let $(M, *)$ be a semigroup and let R be a semiring such that (R, M) is a convolution context. If $m, m' \in M$ and $r, r' \in R$ then $p_{m,r} \langle * \rangle p_{m',r'} = p_{m*m', rr'}$ in R^M. Thus $pt(R^M)$ is closed under the operation $\langle * \rangle$ and so is a subsemigroup of $(R^M, \langle * \rangle)$. Indeed, we have a monomorphism of semigroups $(M, *) \to (R^M, \langle * \rangle)$ given by $m \mapsto p_{m,1}$. If $(M, *)$ is a monoid with unit element e, then $(pt(R^M), \langle * \rangle)$ is a monoid with unit element p_{e1} and $p_{m,r} \in R^M$ is idempotent with respect to $\langle * \rangle$ precisely when m is an idempotent element of M and r is an idempotent element of R. If $(M, *)$ is a group and if R is a semifield, then $(pt(R^M) \setminus \{p_0\}, \langle * \rangle)$ is also a group, in which the inverse of $p_{m,r}$ is just $p_{m^{-1}, r^{-1}}$.

If $(M, *)$ and $(M', \diamond)$ are semigroups [monoids] and if $u \colon M \to M'$ be a semigroup [monoid] homomorphism, then we also have a canonical semigroup [monoid] homomorphism $w_u \colon (pt(R^M), \langle * \rangle) \to (pt(R^{M'}), \langle \diamond \rangle)$ defined by $w_u \colon p_{m,r} \mapsto p_{u(m),r}$.

Recall that $H(m) = \{p_{m,r} \mid r \in R\}$ for each $m \in M$. From the above discussion, it is clear that $H(m) \langle * \rangle H(m') \subseteq H(m * m')$ for all $m, m' \in M$.

(4.1) Proposition. *Let R be a semiring and let $(M, *)$ be a semigroup. If N is a finite subgroup of M satisfying the condition that $r = |N|1_R$ is a multiplicative unit of R, then $\sum_{n \in N} r^{-1} p_{n,1}$ is an element of R idempotent with respect to $\langle * \rangle$.*

PROOF. Set $t = \sum_{n \in N} p_{n,1}$. Then for each $x \in N$ we have

$$p_{x,1} \langle * \rangle t = \sum_{n \in N} p_{x,1} \langle * \rangle p_{n,1} = \sum_{n \in N} p_{x*n,1}.$$

Since N is a group, we have $\sum_{n \in N} p_{x*n,1} = t$ and so

$$t\langle * \rangle t = \sum_{x \in N} p_{x,1}\langle * \rangle t = \sum_{x \in N} t = rt$$

whence $r^{-1}t\langle * \rangle r^{-1}t = r^{-2}(t\langle * \rangle t) = r^{-2}(rt) = r^{-1}t$. □

Let us now look at several examples of this construction.

(I) EXAMPLE. The semiring $\mathbb{B}$ is complete and so for any monoid $(M, *)$ we can consider the convolution algebra $(\mathbb{B}^M, +, \langle * \rangle)$. Any function $f \in \mathbb{B}^M$ can be considered as the characteristic function of the subset $supp(f)$. Moreover,

$$supp(f + g) = supp(f) \cup supp(g)$$

and $supp(f\langle * \rangle g) = \{m' * m'' \mid (m,', m'') \in supp(f) \times supp(g)\}$ so that this convolution algebra is just the semiring of all subsets of the monoid M.

The investigation of the algebra $\mathbb{B}^M$, where M is a group, originates in the work of Frobenius. The elements of $\mathbb{B}^M$ are known as *complexes* and their elementary properties are given in most older textbooks on group theory, such as [256, 424]. Indeed, Zassenhaus [424] defines the notion of a complex when M is an arbitrary semigroup. This algebra was also studied in [375] by Tamura and Shafer, and there they coined the term "power semigroup" to describe it. Also see [52], where this construction is extended to n-ary operations on M. It appears in many situations, for example, in [193] it is used to consider indefinite integrals as multifunctions.

If S is any semiring then an *R-power representation* is a morphism of semirings $\gamma: S \to (R^M, +, \langle * \rangle)$. In Proposition 3.4 we saw that if R is a complete semiring with necessary summation and if G is an additive set of generators for R each member of which is prime then the function $\rho_A: R^A \to (\mathbb{B}^A)^G$ is a monic morphism of semirings. This defines a $\mathbb{B}$-power representation $R^A \to \mathbb{B}^{A \times G}$.

In general, a morphism of semirings from a semiring S to $(\mathbb{B}^{A \times A}, +, \langle * \rangle)$, for some nonempty set A, is an *S-interpretation* of the set A. The set A is the *semantic domain* of this interpretation. Such interpretations are studied in [258]. An obvious generalization of this situation is to consider morphisms of semirings $S \to (R^{A \times A}, +, \langle * \rangle)$, where R is a suitable semiring. Such a morphism of semirings will be called an *S-interpretation* of A over R.

(II) EXAMPLE. Let $R = (\mathbb{I}, \vee, *)$, where $*$ is a triangular norm on $\mathbb{I}$, and let (ρ, λ) be a pair of decreasing continuous functions from $\mathbb{I}$ to itself satisfying the condition that $\rho(0) = \lambda(0) = 1$ and $\rho(1) = \lambda(1) = 0$. Following [87], we say that a $\lambda - \rho$ *fuzzy number* is a function $f_{a,b,c}: \mathbb{R} \to \mathbb{I}$, where a is an arbitrary real number and b, c are positive real numbers, defined by

$$f_{a,b,c}: x \mapsto \begin{cases} \rho(\frac{x-a}{c}) & \text{for all } a \leq x \leq a+c \\ \lambda(\frac{a-x}{b}) & \text{for all } a-b \leq x \leq a \\ 0 & \text{otherwise} \end{cases}$$

Convolutions of such numbers of the form $f_{a,b,c}\langle + \rangle f_{a',b',c'}$ have been studied in detail by Mesiar [278].

(III) EXAMPLE. If M is a monoid, the operation $\langle * \rangle$ is often called the *Cauchy product* and R^M is the *algebra of formal power series* in M over R. This construction is well-known and frequently encountered, under various guises, in the literature. The algebra of formal power series in M over R is usually denoted by $R\langle\!\langle M \rangle\!\rangle$. The subsemiring $R^{(M)}$ is the *monoid semiring* defined by M over R and is sometimes denoted by $R[M]$.

One of the most frequently-encountered examples of this construction occurs when we have a nonempty set A and consider the free monoid $M = A^*$ of all words in A, the neutral element of which is the null word $\square$. In this case it is usual to write $R\langle\!\langle A \rangle\!\rangle$ instead of $R\langle\!\langle A^* \rangle\!\rangle$. In particular, if $A = \{x\}$ then $R\langle\!\langle x \rangle\!\rangle$ is the semiring of formal power series in the indeterminate x. Note that if the set A is finite, say $A = \{a_1, \ldots, a_k\}$, then we have a bijective function $\varphi: A^* \to \mathbb{N}$ defined by $\varphi(\square) = 0$ and

$$\varphi(a_{i_0} a_{i_1} \ldots a_{i_n}) = \sum_{h=0}^{n} i_h k^h.$$

If $R = \mathbb{B}$ then the elements of $\mathbb{B}^M$ are then called *languages* in the *alphabet* A. This power algebra was first studied by Kleene [207] and is now at the heart of algebraic automata theory. Refer, for example, to [346]. The extension of interest from $\mathbb{B}^M$ to R^M for suitable semirings R is due to Eilenberg [98].

A *partial semigroup* is a nonempty set M on which we have an operation $*$ defined on some subset of M satisfying the condition that if $m, m', m'' \in M$ which is associative in the sense that if one side of the equation

$$m * (m' * m'') = (m * m') * m''$$

is defined, so is the other and equality in fact holds. If there exists an identity element $e \in M$ such that $e*m$ and $m*e$ are defined and equal to m for all $m \in M$, then M is a *partial monoid*. If M is a partial semigroup [resp. monoid] and $z \notin M$ then we can extend the operation $*$ to $M \cup \{z\}$ and turn it into a semigroup [resp. monoid] by setting

(1) $m * m' = z$ if $m, m' \in M$ and $m * m'$ is not defined in M;
(2) $m * z = z * m = z$ for all $m \in M \cup \{z\}$.

This construction is encountered in several important instances. For example, if $M = A^*$ as above we can consider a partial operation $\wr$, called the *fusion product*, defined as follows: if $m = ua$ and $m' = av$, where $a \in A$ and $u, v \in A^*$, then $m \wr m' = uav$.

To consider another example, let $\Gamma = (Q_0, Q_1)$ be a directed graph, where Q_0 is a nonempty set of *nodes* and Q_1 is a nonempty set of *arrows* $\alpha = (s_\alpha, e_\alpha) \in Q_0 \times Q_0$. If $m \geq 0$ we define a *path* of length m in Γ to be as follows:

(1) A path of length 0 is just $\square$;
(2) A path of length 1 is an element of Q_1;
(3) A path of length $m > 1$ is a word $\alpha_1 \cdot \ldots \cdot \alpha_m \in Q_1^*$ satisfying the condition that $e_{\alpha_i} = s_{\alpha_{i+1}}$ for all $1 \leq i < m$.

This construction turns Q_1^* into a partial monoid which, as noted, can be extended to a monoid $Q_1^* \cup \{z\}$. It forms the basis for the theory of *quivers* [114], which is important in representation of algebras.

(IV) EXAMPLE. Let $R = (\mathbb{R} \cup \{\infty\}, \wedge, +)$ and let A be the monoid $(\mathbb{R}, +)$. The semiring $(R^A, \wedge, \langle + \rangle)$ has important applications in optimal control theory. See [15, 119] for details. Of similar interest is the case in which $R = (\mathbb{R} \cup \{-\infty, \infty\}, \vee, +)$ and $A = (\mathbb{R}, +)$. For applications of the semiring R^A also refer to [119]. Note that in this semiring, the identity element with respect to the convolution operation $\langle + \rangle$ is the function $f: \mathbb{R} \to R$ defined by

$$f: t \mapsto \begin{cases} -\infty & \text{when } t \neq 0 \\ 0 & \text{otherwise} \end{cases}.$$

(V) EXAMPLE. If $R = (\mathbb{R}^+, +, \cdot)$ and if $M = (\mathbb{N}, +)$ then calculations of convolutions in $(R^M, +, \langle + \rangle)$ can be done in $O(n \log(n))$ time using fast Fourier transform methods [5]. However, the situation is much more complicated if $R = (\mathbb{R} \cup \{\infty\}, \wedge, +)$. One fast algorithm can be found in [54].

(VI) EXAMPLE. Let V be a vector space over a field F and let R be the semiring $(\mathbb{R} \cup \{-\infty, \infty\}, \vee, \wedge)$. The operator $\langle + \rangle$ on R^V is sometimes called *Minkowski addition* and has important applications in mathematical morphology. See [130, 359] for details. The more general situation of morphilogy in R^V, where R is a complete distributive lattice, is studied in [341].

(VII) EXAMPLE. [119] Let $R = (\mathbb{R} \cup \{-\infty\}, \vee, +)$ and let $A = (\mathbb{Z}, +)$. Define the function $f_0 \in R^A$ by

$$f_0: k \mapsto \begin{cases} -\infty & \text{if } k < 0 \\ 0 & \text{otherwise} \end{cases}.$$

Then a function $g \in R^A$ is increasing if and only if $g\langle + \rangle f_0 = g$. Indeed, to see this we note that for any $k \in \mathbb{Z}$ we have

$$(g\langle + \rangle f_0): k \mapsto \bigvee_{h \in \mathbb{Z}} [g(k-h) + f_0(h)] = \bigvee_{h \geq 0} g(k-h) = \bigvee_{h \leq k} g(h).$$

This observation is important in the study of timed Petri nets with daters.

(VIII) EXAMPLE. If $(M, *)$ is a monoid with multiplicative identity e and if R is a simple semiring then the subset S of the power algebra $(R^M, +, \langle * \rangle)$ consisting of all those functions $f \in R^M$ satisfying $f(e) = 1_R$ is a subsemiring of R^M. In the case of $R = \mathbb{I}$ this *normalization condition* is often assumed.

(IX) EXAMPLE. [18, 357] Let $M = \mathbb{R}^+$, and let $R = (\mathbb{I}, \vee, *)$, where $*$ is a lower-continuous triangular norm. A function $d \in \mathbb{I}^M$ is a *distance distribution function* if and only if it is monotone, left continuous, and has supremum 1. The set D of R^M consisting of all distance distribuition functions is a subsemiring of the power algebra $(R^M, \vee, \langle * \rangle)$

(X) EXAMPLE. [428] Let $(H, +)$ and $(\bar{H}, +)$ be disjoint copies of the totally-ordered abelian group $(\mathbb{R}, +)$. Thus there exists an order-preserving isomorphism between H and $\bar{H}$ given by $a \mapsto \bar{a}$. Let $G = H \cup \bar{H}$. For convenience, we will also write $a = \bar{\bar{a}}$ for all $a \in H$ and thus consider the map $a \mapsto \bar{a}$ an involution of G. If $t \in G$ set

$$\|t\| = \begin{cases} t & \text{if } t \in H \\ \bar{t} & \text{if } t \in \bar{H} \end{cases}.$$

Define an operation $\oplus$ on G by setting

(1) $a \oplus b = \bar{a} \oplus \bar{b} = a + b \in H$; and
(2) $a \oplus \bar{b} = \bar{a} + b = \overline{a + b} \in \bar{H}$

for all $a, b \in H$. Then $(G, \oplus)$ is an abelian group with additive identity 0. A function $f \in \mathbb{N}^G$ is *well-ordered* if and only if $\{\|t\| \mid t \in supp(f)\}$ is well-ordered (i.e. each of its subsets has a minimal element). Let U be the set of all well-ordered functions in $\mathbb{N}^G$ having countable support. If $f, g \in U$ then

$$f + g: t \mapsto f(t) + g(t)$$

and

$$f\langle + \rangle g: t \mapsto \sum_{t'+t''=t} f(t')g(t'')$$

are again elements of U and $(U, +, \langle + \rangle)$ is a semiring.

(XI) EXAMPLE. If $(M, *)$ is a partial semigroup, then, as we have already noted, we can simply embed M in a semigroup $M_p = M \cup \{z\}$, where z is not an element of M. Given a semiring R, we can identify R^M with the subset of all elements f of R^{M_p} satisfying $f(p) = 0$.

For example, let A be a nonempty set and let $M = A \times A$. Define an operation $*$ on M by setting $(a, a') * (a', a'') = (a, a'')$ and leaving the operation undefined for all other elements of M. If $f, g \in R^M$ and $(a, a'') \in M$ then

$$(f\langle * \rangle g)(a, a'') = \sum_{a' \in A} f(a, a')g(a', a'').$$

So $f\langle * \rangle g = f \circ g$. We note, in particular, that if $f \in R^M$ is an R-equivalence relation then

$$(f\langle * \rangle f)(a, a'') = \sum_{a' \in A} f(a, a')f(a', a'') \succeq f(a, a)f(a, a'') = f(a, a'')$$

since $f(a, a) = 1$.

Thus, if $A = \{1, \ldots, n\}$ then $(R^M, +, \langle * \rangle)$ is just the semiring of all $n \times n$ matrices over R. If $A = \mathbb{N}$ we can consider the subset S of R^M consisting of all *Jacobi matrices*, i.e. those functions $f \in R^M$ satisfying the conditions that for each $h \in \mathbb{N}$ the sets $\{i \in \mathbb{N} \mid f(i, h) \neq 0_R\}$ and $\{j \in \mathbb{N} \mid f(h, j) \neq 0_R\}$ are both finite. Then $(S, +, \langle * \rangle)$ is a semiring. See [219] for the special case of this construction in which R is taken to be a semiring of formal power series.

(XII) EXAMPLE. Each subset L of a semigroup $(M, *)$ defines a congruence relation $\cong_L$ on M, called the *syntactic congruence* of L, and defined as follows: if $m, m' \in M$ then $m \cong_L m'$ if and only if for all $u, v \in M$ we have $u * m * v \in L$ when and only when $u * m' * v \in L$. See [98] for details. The monoid $M_L = M/\cong_L$ is called the *syntactic monoid* defined by L and the power algebra $R^{(M_L)}$, where R is a suitably-chosen semiring, is called the *syntactic algebra* of L. Also refer to [22].

(XIII) EXAMPLE. In their study of block design theory [160], Graver and Jurkat consider functions in $\mathbb{N}^{\mathbb{P}(V)}$, where V is a finite set. This context presents three different convolution operations: $\langle \cup \rangle$, $\langle \cap \rangle$, and $\langle \Delta \rangle$, where Δ is the symmetric difference operator.

(XIV) EXAMPLE. Chakraborty and Khare [60] used convolutions to generalize the notion of a morphism between semigroups. Let $(M, *)$ and $(N, \diamond)$ be semigroups and let R be a semiring such that (R, N) is a convolution context. Let $f \in R^{M \times N}$ be an R-valued map. For each $m \in M$ and we then have an R-valued subset $f_m \in R^N$ defined by $f_m : n \mapsto f(m, n)$ We say that f is an R-valued homomorphism between M and N if, for all $m, m' \in M$ and all $n \in N$,

$f_{m*m'} = f_m \langle \diamond \rangle f_{m'}$. That is to say, f is an R-valued homomorphism from M to N if and only if
$$f(m*m', n) = \sum_{x \diamond y = n} f(m, x) f(m', y)$$
for all $m, m' \in M$ and all $n \in N$.

If A is an finite nonempty set then we can realize the semiring R^A as a morphic image of a power algebra of the form $(R^M, +, \langle * \rangle)$ as follows: set $M = A \cup \{x\}$ for some element $x \notin A$ and define the operation $*$ on M by
$$m * m' = \begin{cases} m, & \text{if } m = m' \in A \\ x, & \text{otherwise.} \end{cases}$$

Then $(M, *)$ is a finitary semigroup and $(f \langle * \rangle g)(a) = f(a)g(a)$ for all $f, g \in R^M$ and all $a \in A$. Thus we have a morphism of semirings $R^M \to R^A$ given by restriction.

If I is an ideal of a semiring R and if $(M, *)$ is a semigroup then $I^M = \{f \in R^M \mid f(m) \in I \text{ for all } m \in M\}$ is an ideal of the hemiring $(R^M, +, \langle * \rangle)$. Similarly, $I^{(M)} = I^M \cap R^{(M)}$ is an ideal of $R^{(M)}$. In particular, $(0)^M$ is the *augmentation ideal* of R^M. This ideal is trivial if and only if R is zerosumfree.

Several alternative definitions of convolution have also been considered in the literature. Kim [203] has considered the following: if $(M, *)$ is a monoid and R is a semiring such that (R, M) is a convolution context, and if $f, g \in R^M$, define the function $f \langle\!\langle * \rangle\!\rangle g \in R^M$ by setting
$$f \langle\!\langle * \rangle\!\rangle g : m \mapsto \sum f(m_1) g(m_1') f(m_2) g(m_2') \cdot \ldots \cdot f(m_k) g(m_k'),$$
where the sum runs over all ways of writing $m = m_1 * m_1' * m_2 * m_2' * \cdots * m_k * m_k'$, for k a natural number and m_i and m_i' elements of M. We will return to this definition in Chapter 8.

Clearly $f \langle * \rangle g \preceq f \langle\!\langle * \rangle\!\rangle g$, with equality not being true in general. Indeed, let M be the dihedral group D_3 with generating set $\{a, b\}$ subject to the relations $a^3 = e = b^2$ and $ba = a^2 b$. Select $t_0 > t_1 > t_2$ in $\mathbb{I}$ and define $f, g \in \mathbb{I}^M$ by setting
$$f : m \mapsto \begin{cases} t_0 & \text{if } m = e \\ t_1 & \text{if } m = b \\ t_2 & \text{otherwise} \end{cases}$$
and
$$g : m \mapsto \begin{cases} t_0 & \text{if } m \text{ is in the subgroup of } M \text{ generated by } ab \\ t_2 & \text{otherwise} \end{cases}.$$

Then $(f\langle *\rangle g)(a) = t_2$ while $(f\langle\!\langle *\rangle\!\rangle g)(a) = t_1$ so $f\langle *\rangle g \neq f\langle\!\langle *\rangle\!\rangle g$. The operation $\langle\!\langle *\rangle\!\rangle$, moreover, has the disadvantage of not being associative, namely in general, if $f, g, h \in R^M$ then $f\langle\!\langle *\rangle\!\rangle(g\langle\!\langle *\rangle\!\rangle h)$ and $(f\langle\!\langle *\rangle\!\rangle g)\langle\!\langle *\rangle\!\rangle h$ are not equal.

Another variant on the convolution product was introduced by Ray [334]. Let $(M, *)$ be a monoid and let R be a monoid satisfying the condition that (R, M) is a convolution context. For a permutation π of M and $f, g \in R^M$, define $f\langle \pi, *\rangle g \in R^M$ by

$$f\langle \pi, *\rangle g \colon m \mapsto \pi^{-1}\left[\sum_{m'*m''=m} f(\pi(m'))g(\pi(m''))\right].$$

If π is the identity permutation then this, of course, reduces to the usual convolution. The primary application of this construction is for the case that M is a group and π is an automorphism of M.

Yet another variant is the following: let R be a semiring and let A be a nonempty set. Assume that to each $a \in A$ we can associate a relation $h_a \in R^{A \times A}$. Then, given $f, g \in R^A$, we now define $f\langle h\rangle g \in R^A$ by setting

$$f\langle h\rangle g \colon a \mapsto \sum_{(a', a'') \in A \times A} f(a')h_a(a', a'')g(a'').$$

In particular, if $(M, *)$ is a finitary semigroup and if, for each $m \in M$, we define $h_m \in R^{M \times M}$ by setting

$$h_m \colon (m', m'') \mapsto \begin{cases} 1 & \text{if } m = m' * m'' \\ 0 & \text{otherwise} \end{cases}.$$

then $f\langle h\rangle g = f\langle *\rangle g$ for all $f, g \in R^M$.

In this manner, one can also construct convolutions with *weighing kernels*. That is to say, if R is a semiring and $(M, *)$ is a semigroup such that (R, M) is a convolution context, and if $h \colon M \times M \to R$ is an R-valued relation on M satisfying the condition that for all $m, m', m'' \in M$ we have

$$h(m, m')h(m * m', m'') = h(m, m' * m'')h(m', m'')$$

then the operation $\langle *\rangle$ on R^M defined by

$$(f\langle *\rangle g)(m) = \sum_{m'*m''=m} f(m')h(m', m'')g(m'')$$

defines the structure of a hemiring on R^M. If M is a monoid then, in order to turn $(R^M, +, \langle *\rangle)$ into a semiring we need, in addition to the conditions given above, the additional condition that $h(m, 1_M) = h(1_M, m) = 1_R$ for all $m \in M$. This construction was given in the case of rings in [337]. Refer also to [111, 370].

Properties of convolutions. We now look at some basic properties of convolutions.

(4.2) PROPOSITION. *If R is a semiring and $(M, *)$ is a semigroup such that (R, M) is a convolution context then:*
(1) $L_r(f) * L_{r'}(g) \subseteq L_{rr'}(f\langle *\rangle g)$ *for all* $f, g \in R^M$ *and all* $r, r' \in R$;
(2) $ht(f\langle *\rangle g) \preceq ht(f)ht(g)$.

PROOF. (1) If $m \in L_r(f) * L_{r'}(g)$ then there exist $m' \in L_r(f)$ and $m'' \in L_{r'}(g)$ such that $m = m' * m''$. Thus

$$(f\langle *\rangle g)(m) = \sum_{x*y=m} f(x)g(y) \succeq f(m')g(m'') \succeq rr'$$

and so $m \in L_{rr'}(f\langle *\rangle g)$.

(2) By definition,

$$ht(f\langle *\rangle g) = \sum_{m \in M} (f\langle *\rangle g)(m)$$
$$= \sum_{m'*m''=m} f(m')g(m'')$$
$$\preceq \left[\sum_{m' \in M} f(m')\right]\left[\sum_{m'' \in M} g(m'')\right]$$
$$= ht(f)ht(g).$$

□

If R is a semiring and $(M, *)$ is a semigroup then a function $f \in R^M$ is *central* if and only if $im(f)$ is contained in the center of R. If M is commutative and (R, M) is a convolution context then for each central $f \in R^M$ and each $g \in R^M$ we have

$$(f\langle *\rangle g)(m) = \sum_{m'*m''=m} f(m')g(m'') = \sum_{m''*m'=m} g(m'')f(m') = (g\langle *\rangle f)(m).$$

(4.3) PROPOSITION. *Let R be a semiring and let $(M, *)$ be a monoid such that (R, M) is a convolution context. If $f \in R^M$ satisfies the condition that $f\langle *\rangle g = g\langle *\rangle f$ for all $g \in R^M$ then $f(x * m) = f(m * x)$ for every invertible $x \in M$.*

PROOF. Let $x \in M$ be invertible and let $g = p_{x^{-1},1}$. Then

$$(f\langle *\rangle g)(m) = \sum_{y*z=m} f(y)g(z) = f(m * x)g(x^{-1}) = f(m * x)$$

and, similarly, $(g\langle *\rangle f)(m) = f(x * m)$. By hypothesis, these two values are equal. □

(4.4) PROPOSITION. *Let R be a semiring and let $(M, *)$ be a monoid such that (R, M) is a convolution context. If $f, g, h \in R^M$ then*

$$f\langle *\rangle(g + h) = (f\langle *\rangle g) + (f\langle *\rangle h)$$

and

$$(f + g)\langle *\rangle h = (f\langle *\rangle h) + (g\langle *\rangle h).$$

PROOF. This is an immediate consequence of the definition. □

(4.5) PROPOSITION. *Let R be a commutative semiring and let $(M, *)$ be a monoid. Then $f\langle *\rangle g = g\langle *\rangle f$ for all $f, g \in R^{(M)}$ if and only if $p_{m,1}\langle *\rangle p_{m',1} = p_{m',1}\langle *\rangle p_{m,1}$ for all $m, m' \in M$.*

PROOF. Clearly the first condition implies the second. Assume therefore that the second condition holds. If $f, g \in R^{(M)}$ then there exist elements $r_1, \ldots, r_k$ and $r'_1, \ldots, r'_t$ of R and there exist elements $m_1, \ldots, m_k$ and $m'_1, \ldots, m'_t$ of M such that $f = \sum_{i=1}^{k} r_i p_{m_i,1}$ and $g = \sum_{j=1}^{t} r'_j p_{m'_j,1}$. The result then follows from the commutativity of R and the distributivity of convolution over addition in R^M. □

(4.6) PROPOSITION. *Let R be a commutative difference-ordered semiring and let $(M, *)$ be a group such that (R, M) is a convolution context. Then the following conditions on $f \in R^M$ are equivalent:*

(1) $f\langle *\rangle g = g\langle *\rangle f$ for all $g \in R^M$;
(2) $f(x * m) = f(m * x)$ for all $m, x \in M$;
(3) $f(m) = f(x^{-1} * m * x)$ for all $m, x \in M$.

PROOF. (1) ⇒ (2): This follows from Proposition 4.3.
(2) ⇒ (3): Assume (1). If $m, x \in M$ then $f((x^{-1} * m) * x) = f(x * (x^{-1} * m)) = f(m)$, proving (2).
(3) ⇒ (1): If $m \in M$ and $g \in R^M$ then

$$\begin{aligned}(f\langle *\rangle g)(m) &= \sum_{x*y=m} f(x)g(y) \\ &= \sum_{y*(y^{-1}*x*y)} g(y)f(y^{-1} * x * y) \\ &\preceq \sum_{y*z=m} g(y)f(z) \\ &= (g\langle *\rangle f)(m)\end{aligned}$$

while
$$(g\langle *\rangle f)(m) = \sum_{x*y=m} g(x)f(y)$$
$$= \sum_{(x*y*x^{-1})*x=m} f(x*y*x^{-1})g(x)$$
$$\preceq \sum_{z*x=m} f(z)g(x)$$
$$= (f\langle *\rangle g)(m)$$

and so we have (1). □

A subset N of a semigroup $(M, *)$ is *left absorbing* if and only if an element $m \in M$ belongs to N whenever there exists an element n of N satisfying $n*m \in N$. Thus, for example, if N is a subsemigroup of a semigroup $(M, *)$ satisfying the condition that $n * N \subseteq N * n$ for each $n \in N$ then $N' = \{m \in M \mid n*m \in N \text{ for some } n \in N\}$ is a left absorbing subsemigroup of M and, indeed, is the smallest left absorbing subset of M containing N [301].

(4.7) PROPOSITION. *Let R be a semiring and let $(M, *)$ be a semigroup such that (R, M) is a convolution context. Let N be a left absorbing subsemigroup of M and let $u: N \to M$ be the inclusion map. Then for $f \in R^N$ and $g \in R^M$ we have $h_u^{-1}[h_u[f]\langle *\rangle g] = f\langle *\rangle h_u^{-1}[g]$.*

PROOF. Write $g = g_1 + g_2$, where $supp(g_1) \subseteq N$ and $supp(g_2) \subseteq M \setminus N$. Then $h_u^{-1}[g]: n \mapsto g_1(n)$ for all $n \in N$. Set $f' = h_u[f]$. If $m \in supp(g_2)$ and $n \in supp(f')$ then $n * m \notin N$. Therefore, if $x \in M$ we have

$$(f'\langle *\rangle g_2)(x) = \sum_{y*z=x} f(y)g_2(z)$$
$$= \sum\{f(n)g_2(m) \mid n \in N, m \in M \setminus N, \text{ and } n*m = x\}$$

and this implies that $supp(f'\langle *\rangle g_2) \subset M \setminus N$. Thus $h_u^{-1}[f'\langle *\rangle g_2]$ is the 0-map so

$$f\langle *\rangle h_u^{-1}[g] = h_u^{-1}[f'\langle *\rangle g] = h_u^{-1}[f'\langle *\rangle g_1] + h_u^{-1}[f'\langle *\rangle g_2] = h_u^{-1}[f'\langle *\rangle g_1],$$

as desired. □

(4.8) PROPOSITION. *Let $(M, *)$ and $(M', \diamond)$ be groups. Let R be a commutative difference-ordered semiring satisfying the condition that both (R, M) and (R, M') are convolution contexts. If $u: M \to M'$ is a group homomorphism then:*
  (1) *If $g \in R^{M'}$ satisfies the condition that $g\langle \diamond \rangle g' = g'\langle \diamond \rangle g$ for all $g' \in R^{M'}$ then $h_u^{-1}[g]\langle *\rangle f' = f'\langle *\rangle h_u^{-1}[g]$ for all $f' \in R^M$.*
  (2) *If u is an epimorphism and if $f \in R^M$ satisfies the condition that $f\langle *\rangle f' = f'\langle *\rangle f$ for all $f' \in R^M$ then $h_u[f]\langle \diamond \rangle g' = g'\langle \diamond \rangle h_u[f]$ for all $g' \in R^{M'}$.*

PROOF. (1) If $x, m \in M$ then

$$\begin{aligned}
h_u^{-1}[g](x^{-1} * m * x) &= g(u(x^{-1} * m * x)) \\
&= g(u(x)^{-1} \diamond u(m) \diamond u(x)) \\
&= g(u(m)) = h_u^{-1}[g](m)
\end{aligned}$$

and so the result follows from Proposition 4.6.

(2) If $y, m' \in M'$ and if $x \in M$ satisfies $u(x) = y$ then

$$\begin{aligned}
h_y[f](m') &= \sum_{u(m)=m'} f(m) \\
&= \sum_{u(m)=m'} f(x * m * m^{-1}) \\
&= \sum_{u(x*m*x^{-1})=y\diamond m'\diamond y^{-1}} f(x * m * x^{-1}) \\
&\preceq \sum_{u(t)=y\diamond m'\diamond y^{-1}} f(t) \\
&= h_u[f](y \diamond m' \diamond y^{-1}).
\end{aligned}$$

Similarly, $h_u[f](y \diamond m' \diamond y^{-1}) \preceq h_u[f](m')$ and so we have equality. The result then follows from Proposition 4.6. □

(4.9) PROPOSITION. Let R be a semiring and let $(M, *)$ and $(N, \diamond)$ be semigroups such that (R, M) and (R, N) are convolution contexts. Let $u: (M, *) \to (N, \diamond)$ be a morphism of semigroups. Then $h_u^{-1}[g_1 \langle \diamond \rangle g_2] \succeq h_u^{-1}[g_1] \langle * \rangle h_u^{-1}[g_2]$ for all $g_1, g_2 \in R^N$, with equality if u is surjective.

PROOF. If $m \in M$ then

$$\begin{aligned}
h_u^{-1}[g_1 \langle \diamond \rangle g_2](m) &= (g_1 \langle \diamond \rangle g_2)(u(m)) \\
&= \sum_{x \diamond y = u(m)} g_1(u(x)) g_2(u(y)) \\
&\succeq (h_u^{-1}[g_1] \langle * \rangle h_u^{-1}[g_2])(m).
\end{aligned}$$

and it is clear that we have equality if u is surjective. □

(4.10) PROPOSITION. Let R be a QLO-semiring and let $(M, *)$ and $(N, \diamond)$ be monoids. If $u: M \to N$ is a morphism of monoids then $h_u[f\langle * \rangle g] = h_u[f]\langle \diamond \rangle h_u[g]$ for all $f, g \in R^M$.

PROOF. Let $n \in N$. If $n \notin im(h_u)$ then $h_u[f\langle * \rangle g](n) = 0$. Moreover, if $n = n_1 \diamond n_2$, at least one of the n_i is not in $im(h)$ and so

$$(h_u[f] \langle \diamond \rangle h_u[g])(n) = \sum_{n_1 \diamond n_2 = n} h_u[f](n_1) h_u[g](n_2) = 0.$$

Hence we can assume that $n \in im(h_u)$. But in that case, if $n \in N$ then

$$(h_u[f]\langle \diamond \rangle h_u[g])(n) = \sum_{n_1 \diamond n_2 = n} h_u[f](n_1) h_u[g](n_2)$$

$$= \sum_{n_1 \diamond n_2 = n} \left[\sum_{u(m_1) = n_1} f(m_1) \right] \left[\sum_{u(m_2) = n_2} g(m_2) \right]$$

$$= \sum_{u(m_1) \diamond u(m_2) = n} f(m_1) g(m_2)$$

$$= \sum_{u(m_1 * m_2) = n} f(m_1) g(m_2)$$

$$= h_u[f\langle * \rangle g](n).$$

□

We have already noted that if R is a semiring and $(M, *)$ is a semigroup such that (R, M) is a convolution context, then M is isomorphic to a subsemigroup of $(R^M, \langle * \rangle)$ and hence any congruence relation θ on R^M induces a congruence relation θ' on M in a canonical manner. Clearly $(R, M/\theta')$ is also a convolution context and we have a canonical morphism of hemirings $\gamma: R^{M/\theta'} \to R^M/\theta$ defined as follows: if $f \in R^{M/\theta'}$ then there exists a function $g \in R^M$ defined by $g: m \mapsto f(m/\theta')$. Set $\gamma(f) = g/\theta$.

If $(M, *)$ and $(N, \diamond)$ are monoids then so is $(M \times N, \#)$, where the operation $\#$ is defined by $(m_1, n_1) \# (m_2, n_2) = (m_1 * m_2, n_1 \diamond n_2)$ for all $m_1, m_2 \in M$ and $n_1, n_2 \in N$. Moreover, if both M and N are commutative or finitary so is $M \times N$.

In addition, if R is a commutative semiring and if M and N are commutative finitary monoids then there exists a canonical isomorphism of semirings $\gamma: (R^M)^N \to R^{M \times N}$ defined by $\gamma(w): (m, n) \mapsto w(n)(m)$ for all $w \in (R^M)^N$ and $(m, n) \in M \times N$. Indeed, this map is clearly well-defined. If $w \ne w' \in (R^M)^N$ then there exists an element $n_0 \in N$ satisfying $w(n_0) \ne w'(n_0)$, which in turn means that there exists an element $m_0 \in M$ such that $w(n_0)(m_0) \ne w'(n_0)(m_0)$. In other words, $\gamma(w): (m_0, n_0) \ne \gamma(w'): (m_0, n_0)$ and so $\gamma(w) \ne \gamma(w')$. Thus γ is monic. If $f \in R^{M \times N}$ then $f = \gamma(w)$, where $w \in (R^M)^N$ is defined by $w(n): m \mapsto f(m, n)$ for all $m \in M$ and $n \in N$. Thus γ is epic. We are left therefore, to show that γ is a morphism of semirings. And, indeed, if $w, w' \in (R^M)^N$ then for all $m \in M$ and $n \in N$ we have

$$\gamma(w + w')(m, n) = [(w + w')(n)](m)$$
$$= [w(n) + w'(n)](m)$$
$$= w(n)(m) + w'(n)(m)$$
$$= \gamma(w)(m, n) + \gamma(w')(m, n)$$

and

$$\gamma(w\langle\diamond\rangle w')(m,n) = [(w\langle\diamond\rangle w')(n)](m)$$
$$= \left[\sum_{n'\diamond n''=n} w(n')\langle *\rangle w'(n'')\right](m)$$
$$= \sum_{n'\diamond n''=n}\left[\sum_{m'*m''=m} w(n')(m')w'(n'')(m'')\right]$$
$$= \sum_{(m',n')\#(m'',n'')=(m,n)} \gamma(w)(m',n')\gamma(w')(m'',n'')$$
$$= [\gamma(w)\langle\#\rangle\gamma(w')](m,n)$$

which establishes our contention.

Let R be a complete semiring and let $(M,*)$ be a semigroup. The function δ from the power algebra $(R^M, +, \langle *\rangle)$ to R given by $\delta\colon f \mapsto \sum_{m\in M} f(m)$ clearly satisfies $\delta(f+g) = \delta(f) + \delta(g)$. Furthermore,

$$\delta(f\langle *\rangle g) = \sum_{m\in M}\left[\sum_{m'*m''=m} f(m')g(m'')\right]$$
$$= \sum_{m'\in M}\sum_{m''\in M} f(m')g(m'')$$
$$= \left[\sum_{m'\in M} f(m')\right]\left[\sum_{m''\in M} g(m'')\right]$$
$$= \delta(f)\delta(g)$$

and so δ is also a morphism of hemirings, called the *augmentation morphism*. We will denote the restriction of δ to $R^{(M)}$ by δ'. This map is well-defined even if R is not complete and M is not finitary and is, again, a morphism of hemirings.

The morphism $\delta\colon R^M \to R$ defines a congruence relation $\equiv_\delta$ on R^M by setting $f \equiv_\delta g$ if and only if $\delta(f) = \delta(g)$, called the *augmentation congruence*. Again, if R is not complete and M is not finitary we define this congruence only on $R^{(M)}$.

Note that if R is a lattice-ordered semiring and $(M,*)$ is a semigroup then the hemiring R^M need not be lattice-ordered. Indeed, while it is always true that $(f+g)(m) = (f \vee g)(m)$ for all $f,g \in R^M$ and all $m \in M$, since both addition and joins are defined componentwise, it need not be true that $f\langle *\rangle g \leq f \wedge g$ for all $f,g \in R^M$. We do know that if $f \in R^M$ satisfies the condition that $m = m' * m''$ in M implies that $f(m) \geq f(m') \vee f(m'')$ then

$$(f\langle *\rangle g)(m) = \sum_{m'*m''=m} f(m')g(m'') \leq \sum_{m'*m''=m} f(m') \leq f(m)$$

so that if both f and g satisfy this condition we do in fact have

$$(f\langle *\rangle g)(m) \leq (f \wedge g)(m).$$

This suggests that we look at

$$S = \{f \in R^M \mid m = m' * m'' \text{ implies } f(m) \geq f(m') \vee f(m'')\}.$$

If $f, g \in S$ then clearly $f + g \in S$. Now suppose that $m = m_1 * m_2$ in M. If $m_1 = m_1' * m_1''$ then $m = m_1' * m_1'' * m_2$ so

$$\begin{aligned}(f\langle *\rangle g)(m_1) &= \sum_{m_1 = m_1' * m_1''} f(m_1') * g(m_1'') \\ &\leq \sum_{m_1 = m_1' * m_1''} f(m_1') g(m_1'' * m_2) \\ &\leq (f\langle *\rangle g)(m)\end{aligned}$$

and similarly $(f\langle *\rangle g)(m_2) \leq (f\langle *\rangle g)(m)$. Thus $f\langle *\rangle g \in S$ and so S is a lattice-ordered semiring if R^M is a semiring.

If R is a QLO-semiring, if $f \in R^M$ and if $\{g_1 \mid i \in \Omega\} \subseteq R^M$ then for each $m \in M$ so

$$\begin{aligned}\left[f\langle *\rangle \left(\sum_{i \in \Omega} g_i\right)\right](m) &= \sum_{m' * m'' = m} \left[f(m') \sum_{i \in \Omega} g_i(m'')\right] \\ &= \sum_{m' * m'' = m} \left[\sum_{i \in \Omega} f(m') g_i(m'')\right] \\ &= \sum_{i \in \Omega} \left[\sum_{m' * m''} f(m') g_i(m'')\right] \\ &= \sum_{i \in \Omega} (f\langle *\rangle g_i)(m) \\ &= \left[\sum_{i \in \Omega} (f\langle *\rangle g_i)\right](m).\end{aligned}$$

Therefore, in particular, S is quantic lattice-ordered as well.

In particular, if R is a QLO-semiring and $(M, *)$ is a semigroup, and if $f \in R^M$, then for each positive integer k define $f_k \in R^M$ by $f_1 = f$ and $f_k = f_{k-1}\langle *\rangle f$ for $k > 1$. Define the $\langle *\rangle$-*transitive closure* of f to be $f^{(*)} = \sum_{k=1}^{\infty} f_k$. Then

$$f^{(*)}\langle *\rangle f^{(*)} = \sum_{j=1}^{\infty}\sum_{k=1}^{\infty} f_j \langle *\rangle f_k = \sum_{k=2}^{\infty} f_k \preceq f^{(*)}$$

and, indeed, $f + f^{(*)}\langle *\rangle f^{(*)} = f^{(*)}$. If $f_1 \preceq f_2$ in R^M then surely $f_1^{(*)} \preceq f_2^{(*)}$.

(4.11) PROPOSITION. Let R be a semiring and let $(M, *)$ be a monoid with identity element e such that (R, M) is a convolution context. If $f, g \in R^M$ satisfying $f(e) = 1_R = g(e)$ then $f, g \preceq f\langle *\rangle g$.

PROOF. If $m \in M$ then

$$f(m) = f(m)1_R = f(m)g(e) \preceq \sum_{m'*m''=m} f(m')g(m'') = (f\langle *\rangle g)(m)$$

and so $f \preceq f\langle *\rangle g$. A similar proof shows $g \preceq f\langle *\rangle g$. □

The condition that $f(e) = 1_R = g(e)$ in Proposition 4.11, or something akin to it, is necessary, as the following example, due to [168], shows. Let $M = \{e, a, b, c\}$ and define an operation $*$ on M as follows:

$$e = e * e$$
$$a = e * a = a * e = b * a = c * a = c * c$$
$$b = e * b = a * c = b * e = b * b = c * b$$
$$c = e * c = a * a = a * b = b * c = c * e.$$

Then $(M, *)$ is a monoid with identity element e. Let $f, g \in \mathbb{I}^M$ be the functions defined as follows:

$$f(e) = 1; \quad g(e) = \frac{1}{3};$$
$$f(a) = \frac{1}{4}; \quad g(a) = \frac{1}{5};$$
$$f(b) = \frac{1}{5}; \quad g(b) = \frac{1}{6};$$
$$f(c) = \frac{1}{8}; \quad g(c) = \frac{1}{4}.$$

Then $g \preceq f\langle *\rangle g$ and $g \preceq g\langle *\rangle f$ while $f \not\preceq f\langle *\rangle g$ and $f \not\preceq g\langle *\rangle f$.

Convolutions on semimodules. Now assume that $(M, +)$ is a left semimodule over a semiring S and let R be a semiring such that (R, M) is a convolution context. Each element s of S defines a function $h_s: M \to M$ by setting $h_s: m \mapsto sm$.

If $f \in R^M$ and if $s \in S$ then we define the function $sf \in R^M$ by setting $sf = h_s[f]$. That is to say, $sf: m \mapsto \sum_{sm'=m} f(m')$. In particular, if S is a ring and M is a left S-module we set $-f = (-1)f$. That is to say, $-f: m \mapsto -m$ for each $m \in M$.

Note, in particular, if that $f \in R^M$ then $0f$ is not the constant function $m \mapsto 0_R$ but rather

$$0f: m \mapsto \sum_{0m'=m} f(m') = \begin{cases} 0_R & \text{if } m \neq 0_M \\ \sum_{m' \in M} f(m') & \text{otherwise} \end{cases}.$$

(4.12) PROPOSITION. [397] Let R and S be semirings and let $(M,+)$ be a left S-semimodule such that (R,M) is a convolution context. If $s, s' \in S$, if $f, g, h \in R^M$ and if $\{f_i \mid i \in \Omega\} \subseteq R^M$ then

(1) $f \preceq g \Rightarrow sf \preceq sg$;
(2) $s(s'f) = (ss')f$;
(3) $s(f\langle+\rangle g) = sf\langle+\rangle sg$;
(4) $s\left(\sum_{i \in \Omega} f_i\right) = \sum_{i \in \Omega} sf_i$;
(5) $(sf)(sm) \succeq f(m)$ for all $m \in M$;
(6) $g(sm) \succeq f(m)$ for all $m \in M$ if and only if $g \succeq sf$;
(7) $(sf\langle+\rangle s'g)(sm + s'm') \succeq f(m)g(m')$ for all $m, m' \in M$;
(8) $h(sm + s'm') \succeq f(m)g(m')$ for all $m, m' \in M$ if and only if $h \succeq sf\langle+\rangle s'g$.

PROOF.

(1) If $f \preceq g$ then there exists a function $h \in R^M$ such that $f + h = g$. Therefore, for each $m' \in M$ we have $f(m') + h(m') = g(m')$ and so, for $m \in M$,

$$(sg)(m) = \sum_{sm'=m} g(m') = \sum_{sm'=m} f(m') + h(m') \succeq (sf)(m)$$

(2) If $m \in M$ then

$$[s(s'f)](m) = \sum_{sm'=m} s'f(m') = \sum_{s(s'm'')=m} f(m'')$$
$$= \sum_{(ss')m''=m} f(m'') = [(ss')f](m)$$

(3) If $m \in M$ then

$$[s(f\langle+\rangle g)](m) = \sum_{sm'=m} (f\langle+\rangle g)(m')$$
$$= \sum_{sm'=m} \sum_{x+y=m'} f(x)g(y)$$
$$= \sum_{s(x+y)=m} f(x)g(y)$$
$$= \sum_{sx+sy=m} f(x)g(y)$$
$$= \sum_{x'+y'=m} \left[\sum_{sx=x'} f(x)\right]\left[\sum_{sy=y'} g(y)\right]$$
$$= \sum_{x'+y'=m} (sf)(x')(sg)(y')$$
$$= [sf\langle+\rangle sg](m)$$

(4) This is a straightforward consequence of the definition.

(5) If $m \in M$ then $(sf)(sm) = \sum_{sm'=sm} f(sm')f(m') \succeq f(m)$.

(6) If $g(rm) \succeq f(m)$ for all $m \in M$ then

$$(rf)(x) = \sum_{ry=x} f(y) \preceq \sum_{ry=x} g(ry) = g(x)$$

for all $x \in M$ and so $g \succeq rf$. Conversely, if $g \succeq rf$ then $g(rm) \succeq (rf)(rm) \succeq f(m)$ for all $m \in M$.

(7) If $m, m' \in M$ then

$$(sf\langle+\rangle s'g)(sm + s'm') \succeq [(sf)(sm)][s'g)(s'm') \succeq f(m)g(m')$$

(8) If $h(sm + s'm') \succeq f(m)g(m')$ for all $m, m' \in M$ then

$$(sf\langle+\rangle s'g)(z) = \sum_{x+y=z} (sf)(x)(sg)(y)$$

$$= \sum_{x+y=z} \left[\sum_{sx'=x} f(x')\right]\left[\sum_{s'y'=y} g(y')\right]$$

$$= \sum_{sx'+s'y'=z} f(x')g(y')$$

$$\preceq h(z)$$

which shows that $h \succeq sf\langle+\rangle s'g$. Conversely, if $h \succeq sf\langle+\rangle s'g$ then $h(sm+s'm') \succeq (sf\langle+\rangle)s'g)(sm + s'm') \succeq [(sf)(sm)][(s'g)(s'm')] \succeq f(m)g(m')$ for all $m, m' \in M$. □

COROLLARY. Let R and S be semirings and let $(M, +)$ be a left S-semimodule such that (R, M) is a convolution context. If $s, s' \in S$ and $f \in R^M$ then

(1) $sf \preceq f$ if and only if $f(sm) \succeq f(m)$ for all $m \in M$;
(2) $sf\langle+\rangle s'f \preceq f$ if and only if $f(sm + s'm') \succeq f(m)f(m')$ for all $m, m' \in M$.

PROOF. This is a direct consequence of Proposition 4.12. □

(4.13) PROPOSITION. [397] Let R and S be semirings and let $(M, +)$ and $(N, +)$ be left S-semimodules such that (R, M) and (R, N) are convolution contexts. If $s, s' \in S$, if $f, g, \in R^M$ and if $\alpha\colon M \to N$ is a morphism of S-semimodules, then

(1) $h_\alpha[f\langle+\rangle g] = h_\alpha([f]\langle+\rangle h_\alpha[g]$;
(2) $h_\alpha[sf] = sh_\alpha[f]$;
(3) $h_\alpha[sf\langle+\rangle s'g] = sh_\alpha[f] + s'h_\alpha[g]$.

PROOF. (1) The proof here is essentially the same as the proof of Proposition 4.12(3).

(2) If $y \in N$ then

$$h_\alpha[sf](y) = \sum_{\alpha(m)=n} (sf)(m)$$
$$= \sum_{\alpha(m)=n} \sum_{sm'=m} f(m')$$
$$= \sum_{\alpha(sm')=n} f(m')$$
$$= \sum_{s\alpha(m')=n} f(m')$$
$$= \sum_{sn'=n} \sum_{\alpha(m')=n'} f(m')$$
$$= sh_\alpha[f](y)$$

(3) This follows from (1) and (2). □

Of course, one can equally well consider convolutions defined by arbitrary n-ary operations, as was already noted by Zadeh [419, 420, 3421]. Also see [185]. For a general construction in a topos-theoretic context, refer to [369]. Let R be a complete semiring and let A be a nonempty set. Let $*: A^n \to A$ be an n-ary operation on A, where n is a positive integer. Then $*$ induces an n-ary operation $\langle * \rangle$ on R^A defined by convolution:

$$\langle * \rangle (f_1, \ldots, f_n) : a \mapsto \sum_{a=*(a_1,\ldots,a_n)} f_1(a_1) \cdot \ldots \cdot f_n(a_n).$$

(4.14) PROPOSITION. *Let R be a CLO-semiring having an additive set of generators G each member of which is prime. If $*$ is an n-ary operation on a nonempty set A then the function $\rho_A : R^A \to (\mathbb{B}^A)^G$ commutes with $\langle * \rangle$ in the following sense: for each such $r \in G$ the diagram*

$$\begin{array}{ccc} (R^A)^n & \xrightarrow{\langle * \rangle} & R^A \\ \pi_r^n \downarrow & & \downarrow \pi_r \\ (\mathbb{B}^A)^n & \xrightarrow{\langle * \rangle} & \mathbb{B}^A \end{array}$$

commutes.

PROOF. Recall that $\infty = 1$ in a CLO-semiring. Let $f_1, \ldots, f_n \in R^A$, let $r \in G$, and let $a \in A$. Then

$$\rho_A(\langle * \rangle (f_1, \ldots, f_n))(r)(a) = \infty \Leftrightarrow (\langle * \rangle (f_1, \ldots, f_n))(a) \not\leq r$$
$$\Leftrightarrow \sum_{a=*(a_1,\ldots,a_n)} f_1(a_1) \cdot \ldots \cdot f_n(a_n) \not\leq r.$$

Since R is a CLO-semiring, this is true if and only if there exist $a_1, \ldots, a_n \in A$ satisfying $a = *(a_1, \ldots, a_n)$ and $f_i(a_i) \not\leq r$ for all $1 \leq i \leq n$. Thus there exist $a_1, \ldots, a_n \in A$ satisfying $a = *(a_1, \ldots, a_n)$ and $\rho_A(f_i)(r)(a_i) = 1$ for all $1 \leq i \leq n$. This is the same as saying that

$$\sum_{a=*(a_1,\ldots,a_n)} \rho_A(f_1)(r)(a_1) \cdot \ldots \cdot \rho_A(f_n)(r)(a_n) = \infty$$

which, in turn, is equivalent to the condition that

$$\langle * \rangle(\rho_A(f_1), \ldots, \rho_A(f_n))(r)(a) = \langle * \rangle(\rho_A(f_1)(r), \ldots, \rho_A(f_n)(r))(a) = \infty,$$

proving our result. □

We now recall some notions from universal algebra. Suppose that we have a nonempty set V the members of which we will call *variables* and a set P the members of which we will call *operations*. With each $* \in P$ we associate a nonnegative integer $n(*)$ called the *arity* of $*$. An *expression* over V and P consists of a finite string of elements of $V \cup P$ and of the three *auxiliary symbols* left parenthesis, right parenthesis, and comma, defined recursively as follows:

(1) If $v \in V$ then v is an expression;
(2) If $* \in P$ is an operation having arity $n(*)$ and if $E_1, \ldots, E_{n(*)}$ are expressions then $*(E_1, \ldots, E_{n(*)})$ is an expression.

Note that the number of distinct elements of V occurring in a given expression is a positive integer, though of course a given variable can occur many times in the expression. For any finite subset W of V let $exp(W; P)$ denote the set of all expressions over V and P which involve no variables other than those from W (though not all of the variables in W need appear in the expression).

(4.15) PROPOSITION. [Head's Metatheorem] *Let R be a CLO-semiring having an additive set of generators each member of which is prime and let A be a nonempty set on which are defined operations $*_1, \ldots, *_k$, where each $*_i$ has arity $n(*_i)$. Let $V = \{v_1, \ldots, v_m\}$ be a finite set of variables, let $P = \{\langle *_1 \rangle, \ldots, \langle *_k \rangle, +, \cdot\}$ and let $\phi, \psi \in exp(V; P)$. Let $C_1, \ldots, C_m$ be nonempty classes of subsets of R^A closed under projections. Then, if $\odot$ is one of the relations $\leq$, $\geq$, or $=$ then the following conditions are equivalent:*

(1) $\phi(f_1, \ldots, f_m) \odot \psi(f_1, \ldots, f_m)$ *whenever* $f_i \in C_i$ *for all* $1 \leq i \leq m$;
(2) $\phi(f_1, \ldots, f_m) \odot \psi(f_1, \ldots, f_m)$ *whenever* $f_i \in C_i \cap \mathbb{B}^A$ *for all* $1 \leq i \leq m$.

PROOF. Clearly (1) and (2) and so we need only prove the reverse conclusion. Furthermore, it suffices to consider the case that $\odot$ equals $\leq$ since the case it equals $\geq$ follows by symmetry and the case it equals $=$ follows from combining the two previous cases.

Thus, assume (2) and assume that (1) is false. Then there exists an element a of A for which $\phi(f_1, \ldots, f_m)(a) \not\leq \psi(f_1, \ldots, f_m)(a)$. Set $r = \psi(f_1, \ldots, f_m)(a)$. Then

$$\phi(\rho_A(f_1)(r), \ldots, \rho_A(f_m)(r))(a) = \phi(\rho_A(f_1), \ldots, \rho_A(f_m))(r)(a)$$
$$= \rho_A(\phi(f_1, \ldots, f_m))(r)(a) = \infty$$

but

$$\psi(\rho_A(f_1)(r), \ldots, \rho_A(f_m)(r))(a) = \psi(\rho_A(f_1), \ldots, \rho_A(f_m))(r)(a)$$
$$= \rho_A(\psi(f_1, \ldots, f_m))(r)(a) = 0.$$

From this contradiction our result follows. $\square$

As an immediate corollary of this result we have the following:

(4.16) PROPOSITION. *If R is a CLO-semiring having an additive set of generators each member of which is prime and if A is an algebra on which we have defined operations $*_1, \ldots, *_k$ then the algebras $\mathbb{B}^A$ and R^A satisfy the same $(\langle *_1 \rangle, \ldots, \langle *_k \rangle, +, \cdot)$-identities.*

If $\gamma \colon R \to S$ is a morphism of semirings and if A is a nonempty set then, as already noted, we have an induced morphism of semirings $\gamma^A \colon R^A \to S^A$ defined by

$$\gamma^A(f) \colon a \mapsto \gamma(f(a)).$$

If $(M, *)$ is a semiring such that (R, M) and (S, M) are convolution contexts, then we note that

$$\gamma^M(f\langle * \rangle g)(m) = \gamma\left(\sum_{m' * m'' = m} f(m')g(m'')\right)$$
$$= \sum_{m' * m'' = m} \gamma(f(m')g(m''))$$
$$= \sum_{m' * m'' = m} (\gamma^M f)(m')(\gamma^M g)(m'')$$
$$= [\gamma^M(f)\langle * \rangle \gamma^M(g)](m)$$

for all $f, g \in R^M$ and all $m \in M$. Thus $\gamma^M \colon R^M \to S^M$ is a morphism of hemirings.

5. Semiring-valued Subsemigroups and Submonoids

Let R be a semiring and let $(M, *)$ be a semigroup. An *R-valued subsemigroup* of M is a function $f \in R^M$ satisfying the condition that

(1) $f(m * m') \succeq f(m)f(m')$ for all $m, m' \in M$.

Thus, for example, if $a \in R$ satisfies the condition that $a \succeq a^2$ then the constant function $f_a : m \mapsto a$ in R^M is an R-valued subsemigroup of M. If $a \in R$ satisfies $a \succeq a^2$ and if m is an idempotent element of M then $p_{m,a} \in pt(R^M)$ is an R-valued subsemigroup of M.

It is sometimes convenient to identify M itself with the constant function $f_1 : m \mapsto 1_R$, and so every semigroup can be considered as an R-valued semigroup of itself.

If $(M, *)$ is a monoid with identity element e, there are two possibilities given in the literature for defining the notion of an R-valued submonoid of M, each with its advantages and disadvantages. We will distinguish between them in the following manner. If R-valued subsemigroup f of a monoid M satisfies the additional condition

(2) $f(e) \succeq f(m)$ for all $m \in M$

then f is a *facile R-valued submonoid* of M. If f satisfies the condition

(2*) $f(e) = 1_R$

then f is a *regular R-valued submonoid* of M. The constant map $m \mapsto 0_R$ is a trivial example of an R-valued submonoid of M which is facile but not regular. The function $i \mapsto e^i$ from $(\mathbb{N}, +)$ to $(\mathbb{R}^+, +, \cdot)$ is an example of an $\mathbb{R}^+$-valued submonoid of $\mathbb{N}$ which is regular but not facile. Condition (2*), regularity, is assumed by most authors – often implicitly – and it implies condition (2), facility, if the semiring R is simple (for example, if R is a bounded distributive lattice). Of course, an R-valued subsemigroup of a monoid $(M, *)$ may satisfy neither of these conditions.

Pavelka [313] has defined forms of fuzzy logic in the context of lattice-valued subsemigroups of a semigroup M. Gerla [124] has shown that if M is taken to be a free semigroup over a given alphabet then it is possible to establish the axiomatizability of certain classes of lattice-valued subsemigroups of M. Let us look at some further examples

(I) EXAMPLE. Let R be a simple additively-idempotent semiring and let $(M, *)$ be a semigroup. By Proposition 0.3, we see that a sufficient condition for $f \in R^M$ to be an R-valued subsemigroup of M is that it be *subadditive*, namely that it satisfy $f(m * m') \succeq f(m) + f(m')$ for all $m, m' \in M$.

In general, let U be a subsemigroup of an additive semigroup $(M, +)$. A function $f: U \to \mathbb{R}$ is *subadditive* if and only if $f(u + u') \leq f(u) + f(u')$ for all $u, u' \in U$. If we think of f as a function from U to the semiring $R = (\mathbb{R} \cup \{\infty\}, \wedge, +)$, we see that f is subadditive if and only if it is an R-valued subsemigroup of U. The study of subadditive functions on subsemigroups of $\mathbb{R}^n$ lies at the heart of functional analysis; see [184].

Similarly, if A is a nonempty set and if U is a subsemigroup of the semigroup $(\mathbb{P}(A), \cup)$, then a function $f: U \to \mathbb{R}$ is *subadditive* if and only if $f(E \cup E') \leq f(E) + f(E')$ for all $E, E' \in U$. Again, if we think of f as a function from U to the semiring $R = (\mathbb{R} \cup \{\infty\}, \wedge, +)$, we see that f is subadditive if and only if it is an R-valued subsemigroup of U. Thus, for example, Lebesgue measure is a regular R-valued subsemigroup of U, where U is a ring of subsets of a given set.

More generally, if $(U, \vee, \wedge)$ is a lattice and if f is a real-valued submeasure on U in the sense of [340] then f is an regular R-valued subsemigroup of U. Similarly, if $g: U \to \mathbb{I}$ is a *fuzzy set of small elements*, namely a function satisfying the conditions

(1) $g(0) = 1$;
(2) if $v \leq \bigvee_{i=1}^n u_i$ then $f(v) \geq \prod_{i=1}^n g(u_i)$;
(3) if $u_i \geq u_{i+1}$ for $i = 1, 2, \ldots$ and $\bigwedge_{i=1}^\infty u_i = 0$ then $\lim_{i \to \infty} g(u_i) = 1$

then g is a regular $\mathbb{I}$-valued subsemiring of U. These constructions have important applications in the general study of metrizability and measure on lattices and ordered spaces.

(II) EXAMPLE. If R is a ring and if $(M, *)$ is a semigroup then any $f \in R^M$ is an R-valued subsemigroup of M since $a \preceq b$ for all $a, b \in R$. This suggests that in such situations it might be worth replacing the relation "$\preceq$" in the definition, and in the subsequent theory we will develop, by another partial order. For example, if we take $R = \mathbb{R}$, together with its natural total order, and if M is any submonoid of the additive monoid $(\mathbb{R}, +)$ then each nonnegative real number b defines an R-valued regular submonoid of M of the form $m \mapsto b^m$.

(III) EXAMPLE. [332] Let $M = \{r \in \mathbb{R} \mid r > 0\}$, which is a group under ordinary multiplication. Let $f \in \mathbb{I}^M$ be the function defined by

$$f: n \mapsto \begin{cases} 0 & \text{if } r \leq 1 \\ 1 & \text{otherwise} \end{cases}.$$

Then f is an R-valued subsemigroup of M which is not an R-valued submonoid of M since $f(1) = 0$.

(IV) EXAMPLE. Let R be a ring and let $(ideal(R), +)$ be the additive monoid of all ideals of R, together with R itself. The set $R - tors$ of all hereditary torsion theories on the category R-mod of unitary left R-modules is a frame (see [141] for details) and so, in particular, is an additively-idempotent semiring. Let $f: ideal(R) \to R-tors$ be the function which assigns to each element I of $ideal(R)$ the smallest torsion theory $\xi(R/I)$ relative to which it is torsion. By Proposition 30.10 of [141] we know that $f(I+H) = f(I) \wedge f(H)$ and so f is an $(R-tors)$-valued submonoid of $ideal(R)$, which is easily seen to be regular.

(V) EXAMPLE. Let $R = (\mathbb{R}^+, +, \cdot)$ and let k be a positive integer. Let M be the multiplicative group of all $k \times k$ matrices over $\mathbb{R}$ having positive determinant. Then the function $det: A \mapsto |A|$ from M to R is a regular R-valued submonoid of M.

(VI) EXAMPLE. Let U be a nonempty countable set and let $(\mathbb{P}(U), \cup)$ be the monoid of all subsets of U. Let R be the semiring $(\mathbb{N} \cup \{\infty\}, \wedge, +)$. Note that $k \preceq n$ in R if and only if there exists an element $n' \in R$ such that $k \wedge n' = n$, i.e. if and only if $k \geq n$ in the usual ordering of $\mathbb{N} \cup \{\infty\}$. Let $f \in R^{\mathbb{P}(U)}$ be the function which assigns to each subset A of U its cardinality $|A|$. Since $|A \cup B| \leq |A| + |B|$ for all $A, B \in \mathbb{P}(U)$, we have $f(A \cup B) \succeq f(A) + f(B)$ and so f is an R-valued subsemigroup of $\mathbb{P}(U)$. Moreover, $f(\varnothing) = 0$ and this is the identity of R with respect to $+$, so f is in fact a regular R-valued submonoid of $\mathbb{P}(U)$. This construction is related to Kuratowski counting measures and their generalizations; see [63].

(VII) EXAMPLE. Let A be a nonempty set and let R be a complete semiring. Consider the semigroup $M = (R^{A \times A}, \circ)$ of R-valued relations on A. Each $a \in A$ defines a function $e_a: M \to R$ defined by setting $e_a: f \mapsto f(a, a)$. Thus, for all $f, g \in M$ we have

$$e_a(f \circ g) = (f \circ g)(a, a) = \sum_{b \in A} f(a, b) g(b, a) \succeq f(a, a) g(a, a) = e_a(f) e_a(g)$$

and so e_a is an R-valued subsemigroup of M.

(VIII) EXAMPLE. [73] Let $M = \mathbb{Z}/(2) \oplus \mathbb{Z}/(2)$ be the (additive) Klein Four-Group, and let $f \in \mathbb{I}^M$ be the function defined by

$$f: m \mapsto \begin{cases} 1 & \text{if } m = (0,0) \\ 0.5 & \text{if } m = (1,0) \\ 0.6 & \text{if } m = (0,1) \\ 0.7 & \text{if } m = (1,1) \end{cases}.$$

If we consider the semiring $(\mathbb{I}, \vee, *)$ then:

(1) f is an $\mathbb{I}$-valued subsemigroup of M if the product $*$ on $\mathbb{I}$ is given by $a * b = ab$ or $a * b = 0 \vee \{a + b - 1\}$.

(2) f is not an $\mathbb{I}$-valued subsemigroup of M if the product $*$ on $\mathbb{I}$ is given by $a * b = a \wedge b$.

(IX) EXAMPLE. Let R be a semiring and let $(M, *)$ be a semigroup. If $f \in R^M$ is an R-valued subsemigroup of M and if $\gamma: R \to R$ is a morphism of semirings then $\gamma f \in R^M$ is also an R-valued subsemigroup of M. If M is a monoid and f is a facile or regular R-valued submonoid of M then γf is too.

(X) EXAMPLE. [21] Let $(M, *)$ be a monoid with identity element e. Define an equivalence relation $\sim$ on $\mathbb{I}^M$ by setting $f \sim g$ if and only if

$$f(m_1) > f(m_2) \Leftrightarrow g(m_1) > g(m_2) \text{ for all } m_1, m_2 \in M.$$

This is clearly an equivalence relation. If $f \in \mathbb{I}^M$ is a facile $\mathbb{I}$-valued submonoid of M which is not the 0-map then there exists a regular $\mathbb{I}$-valued submonoid g of M satisfying $f \sim g$. Indeed, if $f(e) = t < 1$ just set $g: m \mapsto t^{-1}f(m)$.

(XI) EXAMPLE. Let $M = (\mathbb{N}, \cdot)$ and let $R = (\mathbb{I}, \vee, \wedge)$. Then the function $f \in R^M$ defined by

$$f: m \mapsto \begin{cases} 1 & m = 0 \\ 1/2 & 0 \neq m \in 2\mathbb{N} \\ 1/3 & \text{otherwise.} \end{cases}$$

is an R-valued subsemigroup of M. It is neither regular nor facile.

(XII) EXAMPLE. Let A be a nonempty set and let A^* be the set of all *words* on A, namely finite sequences $w = a_1 a_2 \ldots a_k$ of elements of A, including the null sequence $\square$. Then A^* is a monoid under the operation of concatenation. Let R be a simple difference-ordered semiring. Then any function $f \in R^A$ can be extended to a function $f^* \in R^{A^*}$ by setting $f^*(\square) = 1_R$ and $f^*(a_1 \ldots a_k) = f(a_1) \cdot \ldots \cdot f(a_k)$. Clearly f^* is an R-valued submonoid of A^* which is both regular and facile.

(XIII) EXAMPLE. [342] Let $(M, *)$ be a semigroup and let R be a semiring. Let $\chi_A \in R^M$ be the characteristic function of a nonempty subset A of M. Then A is a subsemigroup of M if and only if χ_A is an R-valued subsemigroup of M.

More generally, let R be a semiring, let $(M, *)$ be a semigroup, and let $f \in R^M$. For each $r \in R$ we have $L_f(r) = \{m \in M \mid f(m) \succeq r\} = \{m \in M \mid p_{m,r} \in pt(f)\}$. If f is an R-valued subsemigroup of M and if $r \in R$ is multiplicatively idempotent then for each $m_1, m_2 \in L_f(r)$ we have $f(m_1 * m_2) \succeq f(m_1)f(m_2) \succeq r^2 = r$ so $m_1 * m_2 \in L_f(r)$. Thus $L_f(r)$, if nonempty, is a subsemigroup of M for each multiplicatively-idempotent $r \in R$. If M is a monoid and f is a facile R-valued submonoid of M then $L_f(r)$, if nonempty, is a submonoid of M.

Conversely, suppose that $1 \succeq r$ for all $r \in R$ that $L_f(r)$ is a subsemigroup of M for each $r \in R$. If $m_1, m_2 \in R$ then $f(m_i) \succeq f(m_1)f(m_2)$ for $i = 1, 2$ and so $m_1, m_2 \in L_f(f(m_1)f(m_2))$. This implies that $m_1 * m_2 \in L_f(f(m_1)f(m_2))$ and so $f(m_1 * m_2) \succeq f(m_1)f(m_2)$, proving that f is an R-valued semigroup of M.

(XIV) EXAMPLE. Let Ω be a nonempty set. For each $i \in \Omega$ let $(M_i, *_i)$ be a monoid and let $M = \coprod_{i \in \Omega} M_i$. That is to say, M consists of all functions $\psi \colon R \to \bigcup_{i \in \Omega} M_i$ satisfying

(1) $\psi(i) \in M_i$ for each $i \in \Omega$, and
(2) $\psi(i) = e_{M_i}$ for all but finitely-many indices $i \in \Omega$.

Then M is a monoid under the operation $*$ defined by $(\psi_1 * \psi_2) \colon i \mapsto \psi_1(i) *_i \psi_2(i)$.

Let R be a commutative semiring and for each $i \in \Omega$ let f_i be an R-valued regular submonoid of M_i. Then the function $f \in R^M$ defined by $f \colon \psi \mapsto \prod_{i \in \Omega} f(\psi(i))$ is an R-valued regular submonoid of M.

(XV) EXAMPLE. Let R be the semiring $(\mathbb{I}, \vee, *)$ for some triangular norm $*$. Menger [275] defined a *statistical metric space* as a set X together with a family of R-valued subsemigroups $\{f_{pq} \mid p, q \in X\}$ of $(\mathbb{R}, +)$ satisfying the following conditions:

(1) $f_{pq}(0) = 0$;
(2) $f_{pq}(r) = 1$ for all $r \in \mathbb{R}$ if and only if $p = q$; and
(3) $f_{pq} = f_{qp}$

for all $p, q \in X$. In that paper, $f_{pq}(x)$ represented the probability that the distance between p and q is less than x.

(XVI) EXAMPLE. Let S be the semiring $(\mathbb{I}, \wedge, *)$, where $*$ is the triangular norm on $\mathbb{I}$ defined by $s * s' = (s + s') \wedge 1$, and let $\delta \colon S \to S$ be the complementation $s \mapsto 1 - s$. If A is a nonempty set then we extend $*$ to R^A by

$$f * g \colon a \mapsto f(a) * g(a)$$

for all $a \in A$. Recall that a tribe in S^A is a subset of S^A closed under $\bar{\delta}$, closed under finite and countably-infinite products, and containing the constant function $f_0: a \mapsto 0$. If T is a tribe in S^A then a *fuzzy game* on T is a function $v \in \mathbb{R}^T$ satisfying $v(f_0) = 0$. Among the more interesting fuzzy games are those which are *superadditive*, namely those which satisfy the condition that $v(f * g) \geq v(f) + v(g)$ for all $f, g \in T$.

If consider fuzzy games as functions from the monoid $(T, *)$ to the semiring $R = (\mathbb{R} \cup \{-\infty\}, \vee, +)$, then the superadditive games are precisely the regular R-valued subsemigroups of $(T, *)$. For a detailed study of fuzzy games and superadditive fuzzy games, refer to [55].

Let $(A, *)$ be a semigroup and let $\diamond$ be the operation on $A \times A$ defined by

$$(a_1, a_2) \diamond (b_1, b_2) = (a_1 * b_1, a_2 * b_2).$$

Then $(A \times A, \diamond)$ is again a semigroup. If R is a semiring and $f \in R^{A \times A}$ is a reflexive R-valued relation on A which is also an R-valued subsemigroup of $(A \times A, \diamond)$, then for all $a, a_1, a_2 \in A$ we have $f(a * a_1, a * a_2) \succeq f(a, a) f(a_1, a_2) = f(a_1, a_2)$ and, similarly, $f(a_1 * a, a_2 * a) \succeq f(a_1, a_2)$. Thus, using the terminology of [245], we see that f is *compatible* with $*$ on A.

(5.1) PROPOSITION. *Let R be a semiring and let $(M, *)$ be a semigroup. If N is a subsemigroup of M then the characteristic function χ_N on N is an R-valued subsemigroup of M. The converse holds if R is zerosumfree.*

PROOF. Assume that N is a subsemigroup of M. If $n_1, n_2 \in N$ then $n_1 * n_2 \in N$ and so $\chi_N(n_1 * n_2) = 1 = \chi_N(n_1) \chi_N(n_2)$. If one of the n_i does not belong to N then $\chi_N(n_1) \chi_N(n_2) = 0$ so clearly $\chi_N(n_1 * n_2) \succeq \chi_N(n_1) \chi_N(n_2)$. Thus χ_N is an R-valued subsemigroup of M.

Now assume that R is zerosumfree and that χ_N is an R-valued subsemigroup of M. If $n_1, n_2 \in N$ then $\chi_N(n_1 * n_2) \succeq \chi_N(n_1) \chi_N(n_2) = 1$, which implies that $\chi_N(n_1 * n_2) \neq 0$. Therefore $\chi_N(n_1 * n_2) = 1$ and so $n_1 * n_2 \in N$, proving that N is a subsemigroup of M. $\square$

Note that if R is a semiring then $p_{e,1}$ is an R-valued regular submonoid of any monoid M and that an R-valued submonoid f of M is regular precisely when $p_{e,1} \preceq f$. Indeed, if f is an R-valued subsemigroup of a semigroup $(M, *)$, recall that

$$pt(f) = \{p_{m,r} \in pt(R^M) \mid f \succeq p_{m,r}\} = \{p_{m,r} \in pt(R^M) \mid f(m) \succeq r\}.$$

We know that $p_{m,r} \langle * \rangle p_{m',r'} = p_{m*m', rr'}$ and so $pt(f)$ is a subsemigroup of the semigroup $(pt(R^M), \langle * \rangle)$. If M is a monoid, then $pt(f)$ is a submonoid of $(pt(R^M), \langle * \rangle)$ precisely when f is regular.

We note that the definition of an R-valued subsemigroup is a "local" one, involving only individual elements. That means that if N is a subsemigroup of a semigroup M and if $f \in R^M$ is an R-valued subsemigroup of M then the restriction of f to N is an R-valued subsemigroup of N.

If S is a semiring and $(M, +)$ is a left S-semimodule then f is an *R-valued S-subsemimodule* of M if, in addition to condition (1) above, it satisfies the conditions

(3) $f(am) \succeq f(m)$ for all $m \in M$ and all $a \in S$;
(4) If $m' + m'' = 0_M$ then $f(m') = f(m'')$.

Note that condition (3) implies that $f(0_M) = f(0_R m) \succeq f(m)$ for all $m \in M$ and so any R-valued left S-subsemimodule of M is also an R-valued facile submonoid of $(M, +)$. Also, if R is additively idempotent and if $a \in S$ has a multiplicative inverse then

$$f(am) \succeq f(m) = f(a^{-1}am) \succeq f(am)$$

and so $f(m) = f(am)$.

Moreover, if R is additively idempotent, condition (1) becomes

(1') $f(m * m') = f(m * m') + f(m)f(m')$

for all $m, m' \in M$, while condition (3) becomes

(3') $f(am) = f(am) + f(m)$

for all $a \in S$ and $m \in M$.

As an example, let S be a commuative semiring and let M be a left S-semimodule. Let $f \colon R \to ideal(S)$ be the function defined by $f \colon m \mapsto (0 : m)$. Then f is an $ideal(S)$-valued subsemimodule of M.

Note too, that if R is a multiplicatively-idempotent semiring and if $f \in R^M$ is an R-valued submonoid of a monoid M then for each $m \in M$ we have $f(m^2) \succeq f(m)f(m) = f(m)$ and so, by induction, $f(m^k) \succeq f(m)$ for all positive integers k.

Properties of R-valued subsemigroups. We now consider various properties of the set of all R-valued subsemigroups of a given semigroup.

(5.2) PROPOSITION. *Let R be a semiring and let $(M, *)$ and $(N, \diamond)$ be monoids. If $f \in R^M$ and $g \in R^N$ are R-valued [facile, regular] submonoids of M and N respectively, one of which is central, then the function $f \times g \in R^{M \times N}$ defined by $f \times g \colon (m, n) \mapsto f(m)g(n)$ is an R-valued [facile, regular] submonoid of $M \times N$.*

PROOF. If $(m_1, n_1), (m_2, n_2) \in M \times N$ then

$$\begin{aligned}(f \times g)((m_1, n_1)(m_2, n_2)) &= (f \times g)((m_1 * m_2, n_1 \diamond n_2)) \\ &= f(m_1 * m_2)g(n_1 \diamond n_2) \succeq f(m_1)f(m_2)g(n_1)g(n_2) \\ &= f(m_1)g(n_1)f(m_2)g(n_2) \\ &= (f \times g)(m_1, n_1)(f \times g)(m_2, n_2).\end{aligned}$$

In the facile case, we also have

$$(f \times g)(e_M, e_N) = f(e_M)g(e_N) \succeq f(m)g(n) = (f \times g)(m, n)$$

for all $(m, n) \in M \times N$. In the regular case, $(f \times g)(e_M, e_N) = f(e_M)g(e_N) = 1_R \cdot 1_R = 1_R$. □

(5.3) PROPOSITION. Let R be a semiring and let $(M, *)$ be a monoid. Let $f \in R^M$ be central. Then f is a regular R-valued submonoid of M if and only if $f \times f \in R^{M \times M}$ is a regular R-valued submonoid of $M \times M$.

PROOF. If f is a regular R-valued submonoid of M then, by Proposition 5.2, we know that $f \times f \in R^{M \times M}$ is a regular R-valued submonoid of $M \times M$. Conversely, suppose this is true. If $m, m' \in M$ then

$$\begin{aligned}f(m * m') &= (f \times f)((m, e) * (m', e)) \succeq (f \times f)(m, e)(f \times f)(m', e) \\ &= f(m)f(e)f(m')f(e) = f(m)f(m')\end{aligned}$$

and so f is an R-valued subsemigroup of M. Moreover, $f(e) = (f \times f)(e, e) = f(e)f(e) = 1_R$ and so f is a regular R-valued submonoid of M. □

(5.4) PROPOSITION. Let R be an entire zerosumfree semiring and let $(M, *)$ be a semigroup. If $f \in R^M$ is a an R-valued subsemigroup of R then $supp(f)$, if nonempty, is a subsemigroup of M. Conversely, if N is a subsemigroup of a semigroup [monoid] M then there exists an R-valued subsemigroup [regular R-valued submonoid] $f \in R^M$ satisfying $supp(f) = N$.

PROOF. Let f be an R-valued subsemigroup of M with $supp(f) \neq \emptyset$ and let $m_1, m_2 \in supp(M)$. Since R is entire, we know that $f(m_1 * m_2) \succeq f(m_1)f(m_2) \neq 0_R$ and so $m_1 * m_2 \in supp(M)$ since R is zerosumfree. Thus $supp(M)$ is a subsemigroup of M.

Conversely, assume that N is a subsemigroup of M and define $f \in R^M$ to be the characeristic function χ_N on N. Then f is an R-valued subsemigroup of M. If M is a monoid and N is a submonoid of M then f is a regular R-valued submonoid of M. □

If R is a semiring and $f, g \in R^M$ for a semigroup $(M, *)$, we have a function $fg \in R^M$ defined by $fg: m \mapsto f(m)g(m)$.

(5.5) PROPOSITION. Let $(R, +, \cdot)$ be a semiring and let $(M, *)$ be a semigroup. If $f, g \in R^M$ are R-valued subsemigroups of M, one of which is central, then fg is an R-valued subsemigroup of M.

PROOF. If $f, g \in R^M$ are R-valued subsemigroups of M then for all $m, m' \in M$ we have

$$(fg)(m * m') = f(m * m')g(m * m')$$
$$\succeq f(m * m')g(m)g(m')$$
$$\succeq f(m)f(m')g(m)g(m')$$
$$= (fg)(m)(fg)(m').$$

If f and g are both facile then $(fg)(e) = f(e)g(e) \succeq f(e)g(m) \succeq f(m)g(m) = (fg)(m)$ for all $m \in M$ and so fg is facile. If f and g are both regular then $(fg)(e) = f(e)g(e) = 1_R \cdot 1_R = 1_R$ and so fg is regular. $\square$

In particular, if $f \in R^M$ is an R-valued subsemigroup of a semigroup $(M, *)$ and if $a \in R$ is central and satisfies $a \succeq a^2$, then $af: m \mapsto af(m)$ is an R-valued subsemigroup of M.

Thus we see that in the situation described in Proposition 5.5, the set of all R-subsemigroups of M is itself a semigroup. Similarly, if R is a semiring and $(M, *)$ is a semigroup then:

(1) If $(M, *)$ is a monoid and $f, g \in R^M$ are facile R-valued submonoids of M one of which is central, then so is fg.

(2) if S is a semiring, if $(M, +)$ is a left S-semimodule, and if $f, g \in R^M$ are R-valued S-subsemimodules of M then so is fg.

Let $(M, *)$ be a semigroup and let R be a semiring. If $f, g \in R^M$ are R-valued subsemigroups of M then $f + g$ need not be an R-valued subsemigroup of M. To see this, let $R = (\mathbb{I}, \vee, \wedge)$ and let $M = F^2$, where F is a field, in which addition is ordinary vector addition. Set $M_1 = \{(a, 0) \mid a \in F\}$ and $M_2 = \{(0, b) \mid b \in F\}$. Choose $t_0 > t_1 > t_2 > t_3$ in $\mathbb{I}$ and define $f, g \in R^M$ as follows:

$$f: v \mapsto \begin{cases} t_0 & \text{if } v \in M_1 \\ t_3 & \text{otherwise} \end{cases}.$$

and

$$g: v \mapsto \begin{cases} t_1 & \text{if } v \in M_2 \\ t_2 & \text{otherwise} \end{cases}.$$

Then f and g are $\mathbb{I}$-valued subsemigroups of M. On the other hand,

$$f \vee g: v \mapsto \begin{cases} t_0 & \text{if } v \in M_1 \\ t_1 & \text{if } (0,0) \neq v \in M_2 \\ t_2 & \text{otherwise} \end{cases}.$$

so $(f \vee g)(1,1) = t_2$ while $(f \vee g)(1,0) \wedge (f \vee g)(0,1) = t_0 \wedge t_1 = t_1 > t_2$.

A nonempty subset U of R^M is *directed* if and only if for each $f, f' \in U$ there exists an $f'' \in U$ satisfying $f \preceq f''$ and $f' \preceq f''$.

(5.6) PROPOSITION. *Let $(M, *)$ be a semigroup and let R be a QLO-semiring. If U is a directed family of R-valued subsemigroups of M then $\sum U$ is also an R-valued subsemigroup of M.*

PROOF. Set $f_0 = \sum U$. If $m, m' \in M$ then

$$\sum_{f \in U} f(m) f(m') \preceq \sum_{f \in U} \sum_{f' \in U} f(m) f'(m').$$

On the other hand, if $f, f' \in U$ there exists an $f'' \in U$ satisfying $f \preceq f''$ and $f' \preceq f''$ so $f(m) f'(m') \preceq \sum_{f'' \in U} f''(m) f''(m')$. Therefore

$$\sum_{f \in U} \sum_{f' \in U} f(m) f'(m') \preceq \sum_{f'' \in U} f''(m) f''(m').$$

Since QLO-semirings are surely difference ordered, this implies that

$$\sum_{f \in U} \sum_{f' \in U} f(m) f'(m') = \sum_{f'' \in U} f''(m) f''(m').$$

In particular, this implies that

$$\begin{aligned}
f_0(m * m') &= \sum_{f'' \in U} f''(m * m') \\
&\succeq \sum_{f'' \in U} f''(m) f''(m') \\
&= \sum_{f \in U} \sum_{f' \in U} f(m) f'(m') \\
&= \sum_{f \in U} f(m) \left[\sum_{f' \in U} f'(m') \right] \\
&= \left[\sum_{f \in U} f(m) \right] \left[\sum_{f \in U} f(m) \right] \\
&= f_0(m) f_0(m')
\end{aligned}$$

and so f_0 is an R-valued subsemigroup of M. □

(5.7) PROPOSITION. *Let R be a complete semiring with necessary summation and let $(M, *)$ be a semigroup. If $f, g \in R^M$ are R-valued subsemigroups of M satisfying $g \preceq f$ then $f\langle * \rangle g \preceq f$ and $g\langle * \rangle f \preceq f$.*

PROOF. If $m \in M$ then

$$(f\langle * \rangle g)(m) = \sum_{m'*m''=m} f(m')g(m'') \preceq \sum_{m'*m''=m} f(m')f(m'') \preceq f(m)$$

so $f\langle * \rangle g \preceq f$. A similar argument shows that $g\langle * \rangle f \preceq f$. □

(5.8) PROPOSITION. *Let R be an additively-idempotent semiring and let $(M, *)$ be a semigroup such that (R, M) is a convolution context where, if R is complete, it is assumed to have necessary summation as well. A function $f \in R^M$ is an R-valued subsemigroup of M if and only if $f\langle * \rangle f \preceq f$.*

PROOF. Assume $f \in R^M$ is an R-valued subsemigroup of M. If $m \in M$ then $f(m) \succeq f(m_1)f(m_2)$ whenever $m_1 * m_2 = m$ and so, in particular,

$$f(m) \succeq \sum_{m_1*m_2=m} f(m_1)f(m_2) = (f\langle * \rangle f)(m),$$

proving that $f \succeq f\langle * \rangle f$. Conversely, assume that $f\langle * \rangle f \preceq f$. If $m_1, m_2 \in M$ then $f(m_1 * m_2) \succeq (f\langle * \rangle f)(m_1 * m_2) \succeq f(m_1)f(m_2)$ and so f is an R-valued subsemigroup of M. □

COROLLARY. *Let R be a CLO-semiring and let $(M, *)$ be a semigroup where, if R is complete, it is assumed to have necessary summation as well. If $f, g \in R^M$ are R-valued subsemigroups of M satisfying $f\langle * \rangle g = g\langle * \rangle f$ then $f\langle * \rangle g$ is an R-valued subsemigroup of M.*

PROOF. Since $f\langle * \rangle g = g\langle * \rangle f$ we have

$$(f\langle * \rangle g)\langle * \rangle (f\langle * \rangle g) = (f\langle * \rangle f)\langle * \rangle (g\langle * \rangle g) \preceq f\langle * \rangle g$$

and so $f\langle * \rangle g$ is an R-valued subsemigroup of M by Proposition 5.8. □

(5.9) PROPOSITION. *Let R be a semiring and let $(M, *)$ be a monoid with identity element e satisfying the condition that (R, M) is a convolution context. Let $f, g \in R^M$ be R-valued submonoids of M, of which f is regular. Then $1 \preceq g(e)$ implies that $f \preceq f\langle * \rangle g$.*

PROOF. If $m \in M$ then

$$(f\langle * \rangle g)(m) = \sum_{m'*m''=m} f(m')g(m'') \succeq f(m)g(e) \succeq f(m)$$

and so $f \preceq f\langle * \rangle g$. □

(5.10) PROPOSITION. Let R be an additively-idempotent semiring with necessary summation and let $(M, *)$ be a monoid such that (R, M) is a convolution context. If f and g are R-valued regular submonoids of M then:

(1) $f\langle*\rangle f = f$; and
(2) $f\langle*\rangle g \succeq f + g$.

PROOF. (1) By Proposition 5.8 we know that $f\langle*\rangle f \preceq f$. Conversely, if $m \in M$ then $(f\langle*\rangle f)(m) = \sum_{m_1 * m_2 = m} f(m_1)f(m_2) \succeq f(m)f(e) = f(m)1_R = f(m)$ so $f\langle*\rangle f \succeq f$, establishing equality.

(2) If $m \in M$ then

$$(f\langle*\rangle g)(m) = \sum_{m'*m''=m} f(m')g(m'') \succeq f(m)g(e) + f(e)g(m)$$
$$= f(m) + g(m) = (f + g)(m).$$

and so $f\langle*\rangle g \succeq f + g$. □

As an example of a situation for which this theorem holds, consider a nonempty set A and let A^* be the monoid of all words on A, the operation being concatenation. A *semi-Thue system* is a nonempty subset T of $A^* \times A^*$, called the *rules* of A^*. Denote the reflexive and transitive closure of T by $\bar{T}$. If $w \in A^*$, set $[w\rangle = \{w' \in A^* \mid (w, w') \in \bar{T}\}$. For $U, V \in \mathbb{P}(A^*)$, set

$$U \diamond V = \bigcup_{w \in U} \bigcup_{w' \in V} [ww'\rangle.$$

Then $R = (\mathbb{P}(A^*), \cup, \diamond)$ is a semiring. Moreover, if $f \in R^{A^*}$ is the function defined by $f: w \mapsto [w\rangle$ then $f\langle\cdot\rangle f = f$.

To see an example of a situation in which the Theorem 5.10 fails if f is not a regular submonoid, consider the following situation, due to [332]. Let $M = (\mathbb{Z}, +)$ and for each positive integer k let $I(k) = \{k, 2k, 3k, \ldots\}$. Then

$$\mathbb{Z} \supset I(2) \supset I(4) \supset \ldots$$

and $\bigcap_{k=1}^{\infty} I(2^k) = \emptyset$. Define a function $f \in \mathbb{I}^M$ as follows:

$$f: n \mapsto \begin{cases} 0 & \text{if } n \in \mathbb{Z} \setminus I(2) \\ 1 - \frac{1}{2^k} & \text{if } n \in I(2^k) \setminus I(2^{k+1}) \end{cases}.$$

Then f is an $\mathbb{I}$-valued subsemigroup of M which is not a regular $\mathbb{I}$-valued submonoid. By Proposition 5.8, this implies that $f\langle+\rangle f \preceq f$. On the other hand, if $h + i = 2$ the then both h and i cannot be in $I(2)$. Therefore

$$(f\langle+\rangle f)(2) = \bigvee_{h+i=2} f(h) \wedge f(i) = 0 < \frac{1}{2} = f(2).$$

so $f\langle +\rangle f \neq f$.

Let $(M,*)$ be a semigroup and let R be a CLO-semiring. Then the set of all $f \in R^M$ satisfying $f\langle *\rangle f = f$ need not be closed under either $\vee$ or $\wedge$. For example [336]: let $R = \mathbb{I}$ and let $M = (\mathbb{R}, \cdot)$. Set $A = \{m \in M \mid m > 1\}$, $B = \{1, 2, 4, 8, \ldots\}$, and $C = (0,1)$ and let χ_A, χ_B, and χ_C be the characteristic functions on A, B, and C respectively. Then $\chi_A\langle\cdot\rangle\chi_A = \chi_A$, $\chi_B\langle\cdot\rangle\chi_B = \chi_B$, and $\chi_C\langle\cdot\rangle\chi_C = \chi_C$, but $\chi_{A\cap B} \neq \chi_{A\cap B}\langle\cdot\rangle\chi_{A\cap B}$ and $\chi_{B\cup C} \neq \chi_{B\cup C}\langle\cdot\rangle\chi_{B\cup C}$.

Consider the following example [213, 347]: Let k be a positive integer and let M be the subspace of the vector space $\mathbb{R}^k$ over $\mathbb{R}$ defined by

$$M = \left\{ (a_1, \ldots, a_n) \in \mathbb{R}^k \,\middle|\, \sum_{i=1}^{k} a_i = 0 \right\}.$$

Let $R = (\mathbb{R} \cup \{\infty\}, \wedge, +)$. The semiring $R^{(M,+)}$ has important applications in multicriteria optimization. Let $f_0 \in R^M$ be the function defined by

$$f_0 \colon (a_1, \ldots, a_n) \mapsto max\{-a_i \mid 1 \leq i \leq n\}.$$

Then $f_0\langle +\rangle f_0 = f_0$ so f_0 is an R-valued submonoid of $(M, +)$. Let S be the subsemiring of R^M consisting of all $g \in R^M$ satisfying $f_0\langle +\rangle g = g\langle +\rangle f_0 = g$. Then S consists of the constant map $m \mapsto \infty$ together with all functions $g \colon M \to R$ satisfying $g(m) - g(m') \leq f_0(m - m')$ for all $m, m' \in M$. Such functions g must be differentiable almost everywhere.

Note that if R is a semiring, if $(M, *)$ is a group such that (R, M) is a convolution context, and if $f, g \in R^M$ are R-valued regular submonoids of M then

$$(f\langle *\rangle g)(e) = \sum_{m \in M} f(m)g(m^{-1}) \succeq f(e)g(e) = 1_R$$

and so, if $1_R \succeq r$ for all $r \in R$, we in fact have $(f\langle *\rangle g)(e) = 1_R$. Thus, if R is also commutative, say if R is a bounded distributive lattice, then the set of all R-valued regular submonoids is a commutive monoid under $\langle *\rangle$.

Let $(M, *)$ be a monoid and let R be a semiring. If α is an endomorphism of M, written as acting on the left, and if $f \in R^M$ then $f\alpha$ is the function in R^M defined by $f\alpha \colon m \mapsto f(\alpha(m))$.

(5.11) PROPOSITION. *Let R be a semiring and $(M, *)$ be a monoid such that (R, M) is a convolution context. If $f, g \in R^M$ and if α is an endomorphism of M*

then $(f\langle *\rangle g)\alpha = f\alpha\langle *\rangle g\alpha$, with equality holding when α is an automorphism of M.

PROOF. If $m \in M$ then

$$[(f\langle *\rangle g)\alpha](m) = \sum_{m'*m''=\alpha(m)} f(m')g(m'')$$
$$\succeq \sum_{x'*x''=m} (f\alpha)(x')(g\alpha)(x'')$$
$$= [(f\alpha)\langle *\rangle(g\alpha)](m)$$

and so $(f\langle *\rangle g)\alpha \succeq (f\alpha)\langle *\rangle(g\alpha)$. If α is an automorphism of M then $x'*x'' = \alpha(m)$ implies that $\alpha^{-1}(x') * \alpha^{-1}(x'') = m$ and so we have equality. $\square$

(5.12) PROPOSITION. Let R be a lattice-ordered semiring and let $(M, *)$ be a monoid. If $f, g \in R^M$ are R-valued [facile, regular] submonoids of M then so is $f \wedge g$. Moreover, if R is a CLO-semiring and if U is a nonempty set of R-valued [facile, regular] submonoids of M then $\wedge U$ is also an R-valued [facile, regular] submonoid of M.

PROOF. If R is a lattice-ordered semiring let $U = \{f, g\}$ be a set of R-valued [facile] submonoids of M. If R is a CLO-semiring let U be an arbitrary nonempty set of R-valued [facile] submonoids of M. Then, for $m, m' \in M$, we have

$$(\wedge U)(m * m') = \wedge\{f(m * m') \mid f \in U\}$$
$$\succeq \wedge\{f(m)f(m') \mid f \in U\}$$
$$\succeq (\wedge U)(m)(\wedge U)(m').$$

The last part of the proposition is immediate. $\square$

Let R be a semiring and let $(M, *)$ be a semigroup such that (R, M) is a convolution context. If $f \in R^M$ then the function $\hat{f} \in R^M$ defined by

$$\hat{f}: m \mapsto \sum_{m=x_1*\cdots*x_t} f(x_1) \cdot \ldots \cdot f(x_t)$$

is an R-valued subsemigroup of M satisfying $f \preceq \hat{f}$. Moreover, if g is an R-valued subsemigroup of M satisfying $f \preceq g$ then it is easy to see that $\hat{f} \preceq g$. We call $\hat{f}$ the R-valued subsemigroup of M generated by f. If R is a CLO-semiring then the set U of all R-valued subsemigroups g of M satisfying $f \preceq g$ for all $f \in A$ is nonempty and, by Proposition 5.12, has a unique minimal element, namely $\wedge U$, and this is precisely $\hat{f}$. Note that if R is a QLO-semiring, then $\hat{f}$ is precisely the $\langle *\rangle$-transitive closure $f^{\langle *\rangle}$ of f which we defined previously.

Let R be a lattice-ordered semiring. If $(M, *)$ is a semigroup and $f, g \in R^M$ are R-valued subsemigroups of M, it does not necessarily follow that $f \vee g$ is a R-valued subsemigroup of M. Indeed, let $R = \mathbb{I}$, let $M = \{a, b, c, d\}$, and let $*$ be the operation on M defined by

$$a = a * a = a * c = b * a = c * a = c * c$$
$$b = a * b = b * b$$
$$c = b * d = d * a = d * c = d * d$$
$$d = a * d = b * c = c * b = c * d = d * b.$$

Define $f, g \in R^M$ by setting

$$f(a) = g(a) = 1$$
$$f(b) = g(c) = \frac{1}{2}$$
$$f(c) = f(d) = g(b) = g(d) = \frac{1}{3}.$$

Then f and g are $\mathbb{I}$-valued subsemigroups of M but $f \vee g$ is not.

(5.13) PROPOSITION. Let R be a semiring. Let $(M, *)$ and $(N, \diamond)$ be semigroups and let $u: M \to N$ be a semigroup homomorphism. If $g \in R^N$ is an R-valued subsemigroup of N then $h_u^{-1}[g]$ is an R-valued subsemigroup of M.

PROOF. Set $f = h_u^{-1}[g] \in R^M$. By definition, $f: m \mapsto g(u(m))$ for all $m \in M$. If $m, m' \in M$ we then have $f(m * m') = g(u(m * m')) = g(u(m) \diamond u(m')) \succeq g(u(m))g(u(m')) = f(m)f(m')$, which is what we needed to show. $\square$

Indeed, if M and N are monoids and if u is a morphism of monoids then $f = h_u^{-1}[g] \in R^M$ is a [facile, regular] R-valued submonoid of M whenever $g \in R^N$ is a [facile, regular] R-valued submonoid of N. We also note that the morphism $w_u: (pt(R^M), \langle * \rangle) \to (pt(R^N), \langle \diamond \rangle)$ restricts to a morphism from $pt(f)$ to $pt(g)$.

(5.14) PROPOSITION. Let R be a QLO-semiring. Let $(M, *)$ and $(N, \diamond)$ be semigroups and let $u: M \to N$ be a semigroup homomorphism. If $f \in R^M$ is an R-valued subsemigroup of M then $h_u[f]$ is an R-valued subsemigroup of N.

PROOF. Set $g = h_u[f] \in R^N$. By definition, $g: n \mapsto \sum_{u(m)=n} f(m)$ for all $n \in N$. We must show that $g(n \diamond n') \succeq g(n)g(n')$ for all $n, n' \in N$. Indeed, if

$n, n' \in N$ then

$$g(n \diamond n') = \sum_{u(m'')=n \diamond n'} f(m'')$$
$$\succeq \sum_{u(m)=n} \sum_{u(m')=n'} f(m * m')$$
$$= \sum_{u(m)=n} \sum_{u(m')=n'} f(m)f(m')$$
$$= \sum_{u(m)=n} \left(f(m) \left[\sum_{u(m')=n'} f(m') \right] \right)$$
$$= \sum_{u(m)=n} f(m)g(n')$$
$$= \left(\sum_{u(m)=n} f(m) \right) g(n')$$
$$= g(n)g(n').$$

□

A nonempty subset N of a semigroup $(M, *)$ is a *left ideal* of M if and only if $m * n \in N$ for all $m \in M$ and $n \in N$; it is a *right ideal* of M if and only if $n * m \in N$ for all $m \in M$ and $n \in N$. A nonzero subset N of M is an *ideal* of M if it is both a left and right ideal of M. Let R be a semiring and let $(M, *)$ be a semigroup. If $f \in R^M$ then

(1) f is an *R-valued left ideal* of M if and only if $f(m * m') \succeq f(m')$ for all $m, m' \in M$;

(2) f is an *R-valued right ideal* of M if and only if $f(m * m') \succeq f(m)$ for all $m, m' \in M$;

(3) f is an *R-valued ideal* of M if and only if it is both an R-valued left ideal and an R-valued right ideal of M.

Note that a nonempty subset N of a semigroup $(M, *)$ is a [left, right] ideal of M if and only if the R-valued characteristic function of N is an R-valued [left, right] ideal of M.

A semigroup $(M, *)$ is an *inversive semigroup* if and only if for each $m \in M$ there exists an element $m' \in M$ satisfying $m * m' * m = m$ and $m' * m * m' = m'$. Let E be the set of all idempotent elements of an inversive semigroup $(M, *)$, partially ordered by the relation $e \preceq f$ if and only if $e * f = e$. Then $f \in \mathbb{I}^M$ is an $\mathbb{I}$-valued right ideal of M if and only if the restriction of f to E is order-reversing and $f(m) = f(m * m')$ for all $m \in M$. See [274] for details.

The following result is based on [231].

(5.15) PROPOSITION. Let R be an complete additively-idempotent semiring with necessary summation and let $(M, *)$ be a semigroup. Let $f_1 \in R^M$ be the constant function $f_1: m \mapsto 1_R$. Then

(1) A function $f \in R^M$ is an R-valued left ideal of M if and only if $f_1 \langle * \rangle f \preceq f$;
(2) A function $f \in R^M$ is an R-valued right ideal of M if and only if $f \langle * \rangle f_1 \preceq f$.

PROOF. (1) First assume that f is an R-valued left ideal of M and let $m \in M$. Then

$$(f_1 \langle * \rangle f)(m) = \sum_{m=m'*m''} f_1(m')f(m'')$$
$$= \sum_{m=m'*m''} f(m'')$$
$$\preceq \sum_{m=m'*m''} f(m)$$
$$= f(m)$$

and so $f_1 \langle * \rangle f \preceq f$.

Conversely, assume that $f_1 \langle * \rangle f \preceq f$. If $m, m' \in M$ then

$$f(m * m') \succeq (f_1 \langle * \rangle f)(m) = \sum_{m=x*y} f_1(x)f(y) \succeq f_1(m)f(m') = f(m')$$

and so f is an R-valued left ideal of M.

(2) This is proven similarly. $\square$

A semigroup $(M, *)$ is *regular* if and only if for each $m \in M$ there exists an $m' \in M$ satisfying $m = m * m' * m$. Iseki [188] showed that a semigroup M is regular if and only if $N_1 N_2 = N_1 \cap N_2$ for each right ideal N_1 and left ideal N_2 of M.

(5.16) PROPOSITION. Let R be a CLO-semiring and let $(M, *)$ be a semigroup. If M is regular then $fg \leq f \langle * \rangle g \leq f \wedge g$ for each R-valued right ideal f and each R-valued left ideal g of M.

PROOF. By Proposition 5.15, we then have $f \langle * \rangle g \leq f \langle * \rangle f_1 \leq f$ and $f \langle * \rangle g \leq f_1 \langle * \rangle g \leq g$ and so $f \langle * \rangle g \leq f \wedge g$. Moreover, if $m \in M$ then

$$(f \langle * \rangle g)(m) = \sum_{m=x*y} f(x)g(y) \geq f(m * m')g(m) \geq f(m)g(m) = (fg)(m)$$

and so $fg \leq f \langle * \rangle g$. $\square$

COROLLARY. *Let R be a complete distributive lattice and let $(M, *)$ be a semigroup. Then M is regular if and only if $f \wedge g = f \langle * \rangle g$ for each R-valued right ideal f and each R-valued left ideal g of M.*

PROOF. If R is regular then, by Proposition 5.16, we see that $f \wedge g = f \langle * \rangle g$ for each R-valued right ideal f and each R-valued left ideal g of M. Conversely, assume this condition holds. Let N_1 be a right ideal of M and let N_2 be a left ideal of M. Clearly $N_1 N_2 \subseteq N_1 \cap N_2$, and so all we have to prove is the reverse inclusion. For each $i = 1, 2$, let $g_i \in R^M$ be the characteristic function on N_i. Then, as already noted, g_1 is an R-valued right ideal of M and g_2 is an R-valued left ideal of M. Moreover, by assumption, $g_1 \langle * \rangle g_2 = g_1 \wedge g_2$ and so, if $m \in N_1 \cap N_2$ then

$$\sum_{m = m' * m''} [g_1(m') \wedge g_2(m'')] = (g_1 \langle * \rangle g_2)(m) = (g_1 \wedge g_2)(m) = 1.$$

This means that there exist elements m' and m'' of M satisfying $m = m' * m''$ and $g_1(m') = 1 = g_2(m'')$. Hence $m = m' * m'' \in N_1 N_2$, proving that $N_1 \cap N_2 \subseteq N_1 N_2$, as desired. □

For the case of $R = \mathbb{I}$, this result is noted in [231].

If $(M, *)$ is a semigroup, if R is a semiring, and $f \in R^M$ is an R-valued ideal of M then surely $f(m' * m * m'') \succeq f(m)$ for all $m, m', m'' \in M$. The converse of this is not true, as the following example, due to [233], shows: let $M = \{m_1, m_2, m_3, m_4\}$ and let $*$ be the operation defined on M as follows:

$$m_i * m_j = \begin{cases} m_2 & \text{if } (i,j) \in \{(3,3), (4,3), (4,4)\} \\ m_1 & \text{otherwise} \end{cases}.$$

Then $(M, *)$ is a semigroup. Let $f \in \mathbb{I}^M$ be the function defined by

$$f(m_i) = \begin{cases} 0.7 & \text{if } i = 1 \\ 0.3 & \text{if } i = 3 \\ 0 & \text{otherwise} \end{cases}.$$

Then $f(m' * m * m'') \succeq f(m)$ for all $m, m', m'' \in M$.but f is not an $\mathbb{I}$-valued ideal of M since $f(m_4 * m_3) = 0 < f(m_3)$.

The following result is based on [230].

(5.17) PROPOSITION. *If R is a multiplicatively-idempotent semiring and if $(M, *)$ is a regular semigroup then the following are equivalent for $f \in R^M$:*

(1) *f is an R-valued ideal of M;*
(2) *$f(m' * m * m'') \succeq f(m)$ for all $m, m', m'' \in M$.*

PROOF. We have already noted that (1) always implies (2). Conversely, assume (2) holds. If $m, m' \in M$ then there exist elements $x, y \in M$ satisfying $m = m*x*m$ and $m' = m'*y*m'$. Then

$$f(m*m') = f((m*x*m)*m') = f((m*x)*m*m') \succeq f(m)$$

by (2). Similary $f(m*m') \succeq f(m')$ so

$$f(m*m') = f(m*m')^2 \succeq f(m) * f(m'),$$

proving (1). □

Morphisms. In the literature, there are several ways of defining morphisms of R-valued semigroups and monoids.

Let M and N be semigroups [resp. monoids] and let R be a semiring. If $f \in R^M$ and $g \in R^N$ are R-valued subsemigroups [resp. submonoids] of M and N respectively, then a morphism of semigroups [resp. monoids] $\alpha: M \to N$ is a morphism *from f to g* if and only if $f \preceq \alpha g$. In this case we write $\alpha: f \to g$ or $\alpha \in Mor(f, g)$. A morphism $\alpha: f \to g$ is an *isomorphism* from f to g if α is an isomorphism of semigroups [resp. monoids]. It is clearly true that if P is another semigroup [resp. monoid] and if $h \in R^P$ then $\beta\alpha \in Mor(f, h)$ whenever $\alpha \in Mor(f, g)$ and $\beta \in Mor(g, h)$.

A related concept is the following: let R be a semiring and let $(M, *)$ and $(N, \diamond)$ be semigroups [resp. monoids]. If $f \in R^M$ and $g \in R^N$ are R-valued subsemigroups [resp. regular submonoids] of M and N respectively, then a morphism of semigroups [resp. monoids] $\alpha: (pt(f), \langle * \rangle) \to (pt(g), \langle \diamond \rangle)$ is called a *point homomorphism* from f to g.

(5.18) PROPOSITION. [102] Let R be a zerosumfree entire semiring and let $(M, *)$ and $(N, \diamond)$ be monoids with identity elements e and e' respectively. Let $f \in R^M$ and $g \in R^N$ be facile R-valued submonoids of M and N respectively and let α be a function from $pt(f)$ to $pt(g)$. Then the following conditions are equivalent:

(1) α is a point homomorphism from f to g;
(2) There exist a morphism of monoids $\varphi: supp(f) \to supp(g)$ and a morphism of multiplicative semigroups

$$\psi: \{r \in R \mid f(e) \succeq r\} \to \{r' \in R \mid g(e') \succeq r'\}$$

such that $\alpha: p_{m,r} \mapsto p_{\varphi(m), \psi(r)}$ for all $m \in supp(f)$ and $r \preceq f(e)$.

PROOF. First note that since R is zerosumfree and entire, by Proposition 5.4 we see that $supp(f)$ and $supp(g)$ are indeed monoids. Note too that if $r, r' \preceq f(e)$ then $p_{e,r}$ and $p_{e,r'}$ belong to $pt(f)$ and so $p_{e,r}\langle * \rangle p_{e,r'} = p_{e,rr'} \in pt(f)$ so $rr' \preceq f(e)$. Thus $\{r \in R \mid f(e) \succeq r\}$ is indeed a multiplicative semigroup, as is $\{r' \in R \mid g(e') \succeq r'\}$.

First let us assume that (1) holds. We define the functions φ and ψ in the following manner:

(a) If $m \in supp(f)$ then $supp(\alpha(p_{m,f(m)})) = \{n\}$ for some unique element n of N. Therefore set $\varphi(m) = n$.

(b) If $r \preceq f(e)$ then there is a unique $r' \in R$ satisfying $\alpha(p_{e,r}) = p_{n',r'}$ for some $n' \in N$. Therefore set $\psi(r) = r'$.

It is now straightforward to establish that φ and ψ satisfy the desired conditions.

Conversely, if (2) holds then it is immediate that α is indeed a point homomorphism. □

Congruences. Let $(M, *)$ be a semigroup and let R be a semiring. An R-valued relation $f \in R^{M \times M}$ is *left compatible* with M if and only if

$$f(m * m', m * m'') \succeq f(m', m'')$$

for all $m, m', m'' \in M$ and it is *right compatible* with M if and only if

$$f(m' * m, m'' * m) \succeq f(m', m'')$$

for all $m, m', m'' \in M$. It is *compatible* with M if it is both left and right compatible with M. An R-valued equivalence relation $f \in R^{M \times M}$ which is compatible with M is an *R-valued congruence relation* on M.

(5.19) PROPOSITION. *Let $(M, *)$ be a semigroup, let R be a semiring, and let $f \in R^{M \times M}$ be an R-valued congruence relation on M. Then the function $g: M \to R$ defined by $g: m \mapsto f(m, 1_R)$ is a regular R-valued subsemigroup of M.*

PROOF. If $m, m' \in M$ then

$$g(m * m') = f(m * m', 1_R) \succeq f(m * m', m') f(m', 1_R)$$
$$\succeq f(m, 1_R) f(m', 1_R) = g(m) g(m')$$

and so g is an R-valued subsemigroup of M, which is clearly regular. □

Now suppose that $(M, *)$ is a semigroup and that $f, g \in R^{M \times M}$ are R-valued equivalence relations on M, at least one of which is central. We have already noted that fg is an R-valued equivalence relation on M. Moreover, if $m, m', m'' \in M$ then

$$(fg)(m * m', m * m'') = f(m * m', m * m'')g(m * m', m * m'')$$
$$\succeq f(m', m'')g(m', m'') = (fg)(m', m'')$$

and similarly $(fg)(m' * m, m'' * m) = (fg)(m'.m'')$ so that fg is an R-valued congruence relation on M as well.

(5.20) PROPOSITION. Let $(M, *)$ be a semigroup, let R be a semiring, and let $f \in R^{M \times M}$ be an R-valued equivalence relation on M. Then f is an R-valued congruence relation if and only if $f(m_1 * m_2, m_3 * m_4) \succeq f(m_1 * m_3)f(m_2 * m_4)$ for all $m_1, m_2, m_3, m_4 \in M$.

PROOF. Assume that f is an R-valued congruence relation and let m_1, m_2, m_3, and m_4 belong to M. Then

$$f(m_1 * m_2, m_3 * m_4) \succeq f(m_1 * m_2, m_3 * m_2)f(m_3 * m_2, m_3 * m_4)$$
$$= f(m_1, m_3)f(m_2, m_4).$$

Conversely, if this condition holds then for all $m, m', m'' \in M$ we have

$$f(m * m'', m' * m'') \succeq f(m, m')f(m'', m'') = f(m, m')$$

and similarly $f(m'' * m, m'' * m') \succeq f(m, m')$, proving that f is an R-valued congruence relation on M. $\square$

If $(M, *)$ is a semigroup then so is $(M \times M, \diamond)$, where $\diamond$ is the operation defined by

$$(m_1, m_3) \diamond (m_2, m_4) = (m_1 * m_2, m_3 * m_4).$$

Thus we see, by Proposition 5.20, that an R-valued equivalence relation on M is an R-valued congruence relation if and only if it is an R-valued subsemigroup of $M \times M$.

Let $(M, *)$ be a semigroup and let R be a semiring. If $\sim$ is a congruence relation on M then the function $f: M \times M \to R$ defined by

$$f: (m, m') \mapsto \begin{cases} 1 & \text{if } m \sim m' \\ 0 & \text{otherwise} \end{cases}.$$

is an R-valued congruence relation on M. Conversely, as in Proposition 2.12, we see that if $f \in R^{M \times M}$ is an R-valued congruence relation on M then the relation $\sim$ on M defined by $m \sim m'$ if and only if $f(m, m') = 1$ is a congruence relation on the semigroup M provided that $\preceq$ is a partial order and $r \preceq 1$ for all $r \in R$.

We see that a semigroup $(M, *)$ is a *metric semigroup* precisely when there exists an $(\mathbb{R} \cup \{\infty\}, \wedge, +)$-valued congruence relation f on M. Kolokol'tsov and Maslov [213] require a stronger condition, namely that f be an $(\mathbb{R} \cup \{\infty\}, \wedge, \vee)$-valued congruence relation; in this case we will say that M is a *strongly metric semigroup*. Similarly, a semiring $(S, +, \cdot)$ is a [strongly] metric semiring if and only if there exists a function $f \colon M \times M$ to $(\mathbb{R} \cup \{\infty\}, \wedge, +)$ [resp. $(\mathbb{R} \cup \{\infty\}, \wedge, \vee)$] such that both with is a congruence relation both on $(S, +)$ and $(S, \cdot)$.

Thus, for example, $(\mathbb{R} \cup \{\infty\}, \wedge, +)$ is a strongly metric semiring under the function $(a, b) \mapsto |e^{-a} - e^{-b}|$. Similarly, $(\mathbb{R} \cup \{-\infty, \infty\}, \wedge, \vee)$ is a strongly metric semiring under the function $(a, b) \mapsto |arctan(a) - arctan(b)|$ and $(\mathbb{R} \cup \{-\infty, \infty\}, \vee, +)$ is a strong metric semiring under the function $(a, b) \mapsto |e^a - e^b|$.

Given a semigroup $(M, *)$ and an $\mathbb{I}$-valued relation $f \in \mathbb{I}^{M \times M}$, Samhan [349] provides a construction for obtaining the smallest $\mathbb{I}$-valued congruence relation on M containing f.

(5.21) PROPOSITION. *Let $(M, *)$ be a semigroup, and let R be a semiring. Let $f \in R^{M \times M}$ be an R-valued congruence and let $r^2 = r \in R$. Then the relation $\cong_{f,r}$ on M defined by $m \cong_{f,r} m'$ if and only if $f(m, m') \succeq r$ is a congruence relation on M.*

PROOF. The relation $\cong_{f,r}$ is clearly reflexive and symmetric. If $m \cong_{f,r} m'$ and $m' \cong_{f,r} m''$ in M then $f(m, m'') \succeq f(m, m')f(m', m'') \succeq r^2 = r$ and so $f(m, m'') \succeq r$, i.e. $m \cong_{f,r} m''$. Thus $\cong_{f,r}$ is an equivalence relation on M. Moreover, if $m, m', x \in M$ and $m \cong_{f,r} m'$ then $f(m * x, m' * x) \succeq f(m, m') \succeq r$ and so $m * x \cong_{f,r} m' * x$. Similarly, $x * m \cong_{f,r} x * m'$. Thus $\cong_{f,r}$ is a congruence relation on M. $\square$

Recall that if $f \in R^{M \times M}$ is an R-valued congruence relation on M and if $m \in M$ then the *congruence class* of m with respect to f is the R-valued function $f_m \in R^M$ defined by $f_m \colon m' \mapsto f(m, m')$. The set of all congruence class of elements of M with respect to f is denoted by M/f.

Now assume that (R, M) is a convolution context. If $m, m' \in M$ then for each

$x \in M$ set $N = \{y * m' \mid y \in M \text{ and } y * z = x\}$. Then we have

$$\begin{aligned}(f_m\langle *\rangle f_{m'})(x) &= \sum_{x=y*z} f_m(y) f_{m'}(z) \\ &= \sum_{x=y*z} f(m,y) f(m',z) \\ &\preceq \sum_{x=y*z} f(m*m', y*m') f(y*m', y*z) \\ &= \sum_{n \in N} f(m*m', n) f(n, x) \\ &\preceq \sum_{n \in M} f(m*m', n) f(n, x) \\ &= (f \circ f)(m*m', x) \\ &\preceq f(m*m', x) \\ &= f_{m*m'}(x)\end{aligned}$$

and so we see that $f_m\langle *\rangle f_{m'} \preceq f_{m*m'}$.

Now assume that R is simple. Following [234], we now define an operation [*] on M/f by setting $f_m[*]f_{m'} = f_{m*m'}$. This is well-defined. Indeed, assume that $f_m = f_{m'}$ and that $f_n = f_{n'}$. We then know that this means that $f(m, m') = f(n, n') = 1$ Therefore

$$\begin{aligned}1 &\succeq f(m*n, m'*n') \\ &\succeq (f \circ f)(m*n, m'*n') \\ &= \sum_{x \in M} f(m*n, x) f(x, m'*n') \\ &\succeq f(m*n, m'*n) f(m'*n, m'*n') \\ &\succeq f(m, m') f(n, n') \\ &= 1 \cdot 1 = 1\end{aligned}$$

and so $f(m*n, m'*n') = 1$, proving that $f_m[*]f_n = f_{m'}[*]f_{n'}$.

Clearly $(M/f, [*])$ is a semigroup and the function $\gamma: M \to M/f$ given by $\gamma: m \mapsto f_m$ is a semiring homomorphism.

(5.22) PROPOSITION. Let $(M, *)$ be a semigroup and let R be a complete simple difference-ordered commutative semiring. If $f, g \in R^{M \times M}$ are R-valued congruence relations on M satisfying $f \circ g = g \circ f$ then $f \circ g$ is an R-valued congruence relation on M.

Moreover, Moreover,

(1) $f \circ g \succeq f$ and $f \circ g \succeq g$;

(2) If h is an R-valued equivalence relation on A satisfying $h \succeq f$ and $h \succeq g$ then $h \circ f \succeq g$.

PROOF. By Proposition 2.4 we know that $f \circ g$ is an R-valued equivalence relation on M. If $m, m', m'' \in M$ then

$$\begin{aligned}(f \circ g)(m', m'') &= \sum_{x \in M} f(m', x) g(x, m'') \\ &\preceq \sum_{x \in M} f(m * m', m * x) g(m * x, m * m'') \\ &\preceq \sum_{y \in M} f(m * m', y) g(y, m * m'') \\ &= (f \circ g)(m * m', m * m'')\end{aligned}$$

and similarly $(f \circ g)(m', m'') \preceq (f \circ g)(m' * m, m'' * m)$. Thus $f \circ g$ is an R-valued congruence relation on M.

The proof of the second part of the proposition is the same as the proof of the second part of Proposition 2.4. □

6. Semiring-valued Subgroups

Let R be a semiring and let $(M, *)$ be a group having identity element e. An R-valued subgroup of M is a R-valued subsemigroup f of M satisfying the additional condition:

(1) $f(m) = f(m^{-1})$ for all $m \in M$;

Clearly, this condition, together with the condition for being an R-valued subsemigroup of M, is equivalent to the condition

(1') $f(m_1 * m_2^{-1}) \succeq f(m_1)f(m_2)$ for all $m_1, m_2 \in M$.

We say that f is *facile* [resp. *regular*] if it is a facile [resp. regular] R-valued submonoid of M.

If f is an R-valued subsemigroup of M satisfying condition (1) and the condition that $f(m)^2 = f(m)$ for all $m \in M$, then

$$f(e) = f(m * m^{-1}) \succeq f(m)f(m^{-1}) = f(m)^2 = f(m)$$

for each $m \in M$ and so f is automatically a R-valued facile subgroup of M. Thus, facility is a very natural condition to consider in the case of multiplicatively-idempotent semirings, such as bounded distributive lattices. Again, we note that if N is a subgroup of a group M and if $f \in R^M$ is an R-valued subgroup of M then the restriction of f to N is an R-valued subgroup of N.

Let $(M, *)$ be a group and let $f \in R^M$ be a facile R-valued subgroup of M. Using a remark in the previous chapter, we see that $L_r(f)$, if nonempty, is a subgroup of M for each multiplicatively-idempotent element r of R. Since $f(m^{-1}) = f(m) \succeq r$ for all $m \in L_r(f)$, we see that $L_r(f)$ is in fact a subgroup of M. If R is a difference-ordered multiplicatively-idempotent semiring and if m and m' are elements of a group $(M, *)$ which are contained in precisely the same subgroups of M, then $f(m) = f(m')$ for every facile R-valued subgroup f of M. Indeed, if f is a facile R-valued subgroup of M then $m \in L_{f(m)}(f)$ implies $m' \in L_{f(m)}(f)$ and so $f(m') \succeq f(m)$. Similarly $f(m) \succeq f(m')$ and so, by difference order, we have equality.

There is an extensive literature on fuzzy subgroups, i.e., $\mathbb{I}$-valued subgroups of a group. See [41] for a study of $int(\mathbb{I})$-valued groups.

Let us now look at some other examples.

(I) EXAMPLE. If $(M, *)$ is a group having identity element e and if R is a semiring then the point p_{e1} is an R-valued subgroup of M which is both facile and regular. Rephrasing a result in [201], we note that the group M is abelian if and only if $f(a^{-1} * b^{-1} * a * b) = 1_R$ for each R-valued regular subgroup f of M. Indeed, if M is abelian then this condition surely holds, by normality. Conversely, if this condition holds then, in particular, $p_{e1}(a^{-1} * b^{-1} * a * b) = 1_R$ and so $a^{-1} * b^{-1} * a * b = e$, proving M abelian.

(II) EXAMPLE. [332] Let $M = \mathbb{R}^{[-1,1]}$, which is a group under addition of functions. For each $0 < r \leq 1$, let

$$N(r) = \{\varphi \in M \mid \varphi(x) \geq 0 \text{ for } -1 \leq x < 0; \varphi(x) = 0 \text{ for } 0 \leq x \leq r\}.$$

Then each $N(r)$ is a subsemigroup of $(M, +)$, but not a subgroup. Let $f \in \mathbb{I}^M$ be the function defined as follows:

$$f: \varphi \mapsto \begin{cases} 1 & \text{if } \varphi \in N(1) \\ r & \text{if } \varphi \in N(r) \setminus \bigcup_{r < s \leq 1} N(s) \\ 0 & \text{otherwise} \end{cases}.$$

Then f is a facile $\mathbb{I}$-valued subsemigroup of R which is not an $\mathbb{I}$-valued subgroup of R.

(III) EXAMPLE. Let $(M, *)$ be a group having identity element e and let R be a semiring. If $f \in R^M$ is a [facile, regular] R-valued subgroup of M and if $m_0 \in M$ then the function $f' \in R^M$ defined by $f': m \mapsto f(m_0^{-1} * m)$ is not necesarily an R-valued subgroup of M. If $f(m_0) \succeq 1_R$ then for all $m', m'' \in M$ we have

$$\begin{aligned}
f'(m' * m'') &= f(m_0^{-1} * m' * m'') \\
&= f(m_0^{-1} * m' * m_0 * m_0^{-1} * m'') \\
&\succeq f(m_0^{-1} * m') f(m_0) f(m_0^{-1} * m'') \\
&\succeq f'(m') f'(m'')
\end{aligned}$$

and so, in this case, it is at least an R-valued subsemigroup of M. The same is true for the function $f'': m \mapsto f(m * m_0^{-1})$. On the other hand, the function $g \in R^M$ defined by $g: m \mapsto f(m_0^{-1} * m * m_0)$ is always a [facile, regular] R-valued subgroup of M, which is said to be *conjugate* to f by m_0.

(IV) EXAMPLE. Let $(V, +)$ be a vector space over $\mathbb{R}$ or $\mathbb{C}$ and let R equal $(\mathbb{R}^+ \cup \{\infty\}, \wedge, +)$. An R-valued subgroup of V is an *(extended) seminorm* on V precisely when $f(av) = |a|f(v)$ for each $v \in V$ and each scalar a. Normed spaces, of course, play an extremely important role in modern analysis.

(V) EXAMPLE. [103] Let R be a semiring and let $r_2 \prec r_1$ be elements of R. If $(M, *)$ is an arbitrary group with identity element e then the function $f \in R^M$ defined by

$$f: m \mapsto \begin{cases} r_1 & \text{if } m = e \\ r_2 & \text{otherwise} \end{cases}$$

is an R-valued subgroup of M, which is not regular unless $r_1 = 1$. Moreover, if M is a cyclic group of prime order then any $\mathbb{I}$-valued of M must be of this form.

The above example can be generalized, using a result based on [354].

(6.1) PROPOSITION. *Let $(M, *)$ be a group having identity element e and let R be a semiring. Let $N_1 \subset \cdots \subset N_t = M$ be a chain of subgroups of M and let $1_R \succeq r_1 \succeq \cdots \succeq r_t$ be elements of R. Then the function $f \in R^M$ defined by*

$$f: m \mapsto \begin{cases} r_1 & \text{if } m \in N_1 \\ r_i & \text{if } m \in N_i \setminus N_{i-1} \text{ for all } 2 \leq i \leq t \end{cases}$$

is a facile R-valued subgroup of M which is regular if and only if $r_1 = 1_R$.

PROOF. For notational convenience, set $N_0 = \emptyset$. Let $m, m' \in M$ be elements satisfying $m \in N_i \setminus N_{i-1}$ and $m' \in N_j \setminus N_{j-1}$. Without loss of generality, we can assume that $i \leq j$. Then $m * m' \in N_j \setminus N_{j-1}$ and so $f(m * m') = r_j \succeq r_i r_j = f(m)f(m')$. Moreover, it is clear that $m^{-1} \in N_i \setminus N_{i-1}$ and so $f(m) = f(m^{-1})$. Finally, $e \in N_1$ and so $f(e) \succeq f(m)$ for all $m \in M$. Thus f is a facile R-valued subgroup of M. The last statement is immediate. $\square$

Now let us return to our examples.

(VI) EXAMPLE. [309] Let R be a commutative semiring and let T be a nonempty subset of R containing an element t_0 which satisfies the condition that $t_0 \succeq t_1 \cdot \ldots \cdot t_k$ for any $\{t_1, \ldots, t_k\} \subseteq T$. Let M be the family of all finite subsets of T. Then $(M, \triangle)$ is a group, where $\triangle$ is the symmetric difference operator: $Y_1 \triangle Y_2 = (Y_1 \cup Y_2) \setminus (Y_1 \cap Y_2)$. Moreover, the function $f \in R^M$ defined by

$$f: Y \mapsto \begin{cases} t_1 \cdot \ldots \cdot t_k & \text{if } Y = \{t_1, \ldots, t_k\} \\ t_0 & \text{if } Y = \emptyset \end{cases}.$$

is a facile R-valued subgroup of M the image of which is the set of all products of finite subsets of elements of T. In particular, if T is a subsemigroup of $(R, \cdot)$ then $\text{im}(f) = T$.

(VII) EXAMPLE. Let A be a nonempty set of symbols and let $A^{-1} = \{a^{-1} \mid a \in A\}$ be a set disjoint from A. Set $B = A \cup A^{-1}$ and let B^* be the set of all finite words on B, i.e. the set of all finite sequences of symbols $w = b_1 b_2 \ldots b_k$, where the $b_i \in B$ (including the nul sequence). Then B is a monoid, the operation on which is concatenation and the identity element of which is the null sequence. Define an equivalence relation $\sim$ on B by setting $w_1 \sim w_2$ if and only if w_1 can be transformed into w_2 by a finite sequence of insertions or deletions of subwords of the form aa^{-1} or $a^{-1}a$ for some $a \in A$. We denote the equivalence class of w with respect to this relation by $[w]$ and we denote the set $B^*/\sim$ of all such equivalence classes by $F(A)$. The operation on B induces an operation on $F(A)$ and it is straightforward to verify that $F(A)$ is in fact a group with respect to this operation, called the *free group* on A. It is well-known that every group is a homomorphic image of a free group.

Now let R be a complete difference-ordered semiring and let $f \in R^A$. Then f can be extended to a function in R^B by setting $f(a^{-1}) = f(a)$ for all $a \in A$. As we saw in the previous chapter, this function can be further extended to a function $f^* \in R^{B^*}$. We define a function $\bar{f} \in R^{F(A)}$ by setting

$$\bar{f} : [w] \mapsto \sum_{w' \sim w} f(w').$$

and it is easily verified that $\bar{f}$ is an R-valued facile subgroup of $F(A)$.

(VIII) EXAMPLE. An R-valued facile or regular submonoid of a group need not automatically be an R-valued facile subgroup. To see this, let M be the group $(\mathbb{Z}, +)$, let $R = \mathbb{R}^+$, and let $f \in R^M$ be the function $f : a \mapsto e^a$. Then f is an R-valued regular submonoid of M but not an R-valued facile subgroup. Another example is given in [161]: let M be as above, let R be any semiring, and let $g \in R^M$ be defined by

$$g : m \mapsto \begin{cases} 0_R & \text{if } m < 0; \\ 1_R & \text{otherwise} \end{cases}.$$

(IX) EXAMPLE. Let R be a semiring and let $(M, *)$ and $(N, \diamond)$ be groups. If $f \in R^M$ and $f \in R^N$ are R-valued facile subgroups of M and N respectively, one of which is central, then we have seen that the function $f \times g \in R^{M \times N}$ defined by

$$f \times g : (m, n) \mapsto f(m)g(n)$$

is an R-valued subsemigroup of $M \times N$. If $(m, n) \in M \times N$ then $(m, n)^{-1} = (m^{-1}, n^{-1})$ and so we see

$$(f \times g)(m, n) = f(m)g(n) = f(m^{-1})g(n^{-1}) = (f \times g)((m, n)^{-1}).$$

Similarly, If $m \in M$ and $n \in N$ then $(f \times g)(e,e) = f(e)g(e) \succeq f(m)g(n) = (f \times g)(m,n)$ and so $f \times g$ is an R-valued facile subgroup of $M \times N$. Refer to [37] for the case $R = \mathbb{I}$.

It is possible, however, for $f \times g$ to be a facile R-valued subgroup of $M \times N$ even though both f and g are not. For example [262], let $(M, *)$ be any group having at least two elements and let $0 \leq h \leq i < 1$ in $\mathbb{I}$. Then the constant function $f: m \mapsto h$ is a $\mathbb{I}$-valued subgroup of M. The function $g \in \mathbb{I}^S$ defined by

$$g: m \mapsto \begin{cases} i & \text{if } m = e \\ 1 & \text{otherwise} \end{cases}$$

is not an $\mathbb{I}$-valued subgroup of M since $g(e) < g(m)$ for all $0 \neq m \in M$. On the other hand, $f \times g = f$ and so $f \times g$ is an $\mathbb{I}$-valued subgroup of $M \times M$.

(X) EXAMPLE. [48] Let $R = (\mathbb{R} \cup \{\infty\}, \wedge, +)$ and let $(M, +)$ be a metrizable abelian group. If d is an invariant metric on M then d is a regular R-valued subgroup of M satisfying the condition that $d(m) \neq 1_R$ for all $m \neq 0_M$. Conversely, if $(M, +)$ is an abelian group and if $f \in R^M$ is a regular R-valued subgroup of M satifying the condition that $f(m) \neq 1_R$ for $m \neq 0_M$ then the function $(m, m') \mapsto f(m - m')$ is an invariant metric of M compatible with its group structure.

Properties of R-valued subgroups. We now look at some elementary properties of R-valued subgroups of groups.

(6.2) PROPOSITION. *Let R be an entire zerosumfree semiring and let $(M, *)$ be a group. If $f \in R^M$ is an R-valued subgroup of M then $supp(f)$ is a subgroup of M. Conversely, if N is a subgroup of M then there exists a regular R-valued subgroup f of M satisfying $supp(f) = N$.*

PROOF. This follows immediately from Proposition 5.4. □

(6.3) PROPOSITION. [308] *Let R be a commutative semiring and let A be a nonempty subset of R which is closed under multiplication and contains an element r_0 satisfying $r_0 \succeq r$ for all $r \in A$. Then there exists an abelian group $(M, *)$ having an R-valued facile subgroup f which satisfies $im(f) = A$.*

PROOF. Let M be the set of all finite subsets of A (including the empty subset) and let $\triangle$ be the symmetric-difference operation. Then $(M, \triangle)$ is a group, the identity element of which is $\varnothing$ and in which each element is its own inverse. Now define $f \in R^M$ as follows:

$$f: B \mapsto \begin{cases} r_0 & \text{if } B = \varnothing \\ r_1 \cdot \ldots \cdot r_k & \text{if } B = \{r_1, \ldots, r_k\} \end{cases}.$$

Then $B \triangle C \subseteq B \cup C$ and so $f(B \triangle C) \succeq f(B)f(C)$. The rest of the conditions follow immediately. □

Let R be a complete semiring, let $(M, *)$ be a group, and let $f \in R^M$ be an R-valued subgroup of M. If A is a nonempty subset of M it is not necessarily true that $\sum_{a \in A} f(a) \in im(f)$, even when the conditions of Proposition 6.3 are satisfied. The following example is due to [9]: Choose $M = (\mathbb{Z}, +)$ and let $f \in \mathbb{I}^M$ be defined by:

$$f: m \mapsto \begin{cases} \frac{1}{2} & \text{if } m = 0 \\ 0 & \text{if } m \text{ is odd} \\ \frac{1}{2}\left(1 - \frac{1}{2^n}\right) & \text{if } m \neq 0 \text{ is divisible by } 2^n \text{ but not by } 2^{n+1} \end{cases}$$

Then f is an $\mathbb{I}$-valued subgroup of of M but

$$1 = \sum_{m \in M} f(m) \notin im(f).$$

(6.4) PROPOSITION. *Let R be a semiring and let $(M, *)$ be a group. If $m_1, m_2 \in M$ and if f is an R-valued subgroup of M satisfying the condition that $f(m_1)f(m_2) \succeq f(m_2)$ then $f(m_1 * m_2) = f(m_1 * m_2)f(m_1)$.*

PROOF. By hypothesis, we know that $f(m_1 * m_2) \succeq f(m_1)f(m_2) \succeq f(m_2)$. So

$$f(m_2) = f(m_1^{-1} * m_1 * m_2) \succeq f(m_1^{-1})f(m_1 * m_2)$$
$$= f(m_1)f(m_1 * m_2) \succeq f(m_1)f(m_2) \succeq f(m_2)$$

and so we have equality. □

(6.5) PROPOSITION. *Let R be a multiplicatively-idempotent semiring and let $(M, *)$ be a group. If $m_1, m_2 \in M$ and if f is an R-valued subgroup of M satisfying $f(m_1) \succeq f(m_2)$ then $f(m_2) = f(m_1)f(m_2)$.*

PROOF. Since $f(m_1) \succeq f(m_2)$ we have $f(m_1)f(m_2) \succeq f(m_2)^2 = f(m_2)$ and so

$$f(m_2) = f(m_1^{-1} * [m_1 * m_2]) \succeq f(m_1^{-1})f(m_1 * m_2)$$
$$\succeq f(m_1)f(m_1)f(m_2) = f(m_1)f(m_2) \succeq f(m_2)$$

and so $f(m_2) = f(m_1)f(m_2)$. □

(6.6) PROPOSITION. *Let R be a difference-ordered semiring and let $(M, *)$ be a group. Let $f \in R^M$ be an R-valued subgroup of M and let $m_1, m_2 \in M$ be elements satisfying the following conditions:*

(1) *$f(m_1)$ is prime in R; and*
(2) *$f(m_1) = f(m_1)f(m_2)$.*

Then either $f(m_1) \succeq f(m_2)$ or $f(m_1 * m_2) = f(m_1)$.

PROOF. We know that $f(m_1 * m_2) \succeq f(m_1)f(m_2) = f(m_1)$. Assume that $f(m_1) \not\succeq f(m_2)$. Then $f(m_1) = f([m_1 * m_2] * m_2^{-1}) \succeq f(m_1 * m_2)f(m_2^{-1}) = f(m_1 * m_2)f(m_2)$ and so by primeness we have $f(m_1) \succeq f(m_1 * m_2)$, establishing equality. $\square$

(6.7) PROPOSITION. Let R be a semiring. Let $(M, *)$ and $(N, \diamond)$ be groups and let $u: M \to N$ be a group homomorphism. If $g \in R^N$ is a [facile, regular] R-valued subgroup of N then $h_u^{-1}[g]$ is a [facile, regular] R-valued subgroup of M.

PROOF. By Proposition 5.13 we know that $f = h_u^{-1}[g]$ is an R-valued submonoid of M. If $m \in M$ then $f(m) = g(u(m)) = g(u(m)^{-1}) = g(u(m^{-1})) = f(m^{-1})$ and so f is in fact an R-valued subgroup of M. $\square$

(6.8) PROPOSITION. Let R be a QLO-semiring. Let $(M, *)$ and $(N, \diamond)$ be groups and let $u: M \to N$ be a group homomorphism. If $f \in R^M$ is a [facile, regular] R-valued submonoid of M then $h_u[f]$ is a [facile, regular] R-valued subgroup of N.

PROOF. By Proposition 5.14 we know that $g = h_u[f]$ is an R-valued submonoid of N. If $n \in N$ then

$$\begin{aligned} g(n^{-1}) &= \sum_{u(m')=n^{-1}} f(m') \\ &= \sum_{u(m^{-1})=n^{-1}} f(m^{-1}) \\ &= \sum_{u(m)=n} f(m^{-1}) \\ &= \sum_{u(m)=n} f(m) \\ &= g(n). \end{aligned}$$

Thus g is an R-valued [facile, regular] subgroup of N. $\square$

(6.9) PROPOSITION. Let R be an additively-idempotent semiring. Let $(M, *)$ and $(N, \diamond)$ be finite groups and let $u: M \to N$ be a group epimorphism. Then there exists a bijective correspondence between the set of all [facile, regular] R-valued subgroups of N and the set of all u-stable [facile, regular] R-valued subgroups of M.

PROOF. This is a direct consequence of Propositions 6.7 and 6.8. $\square$

If $(M, *)$ is a group and R is a semiring then the sum of facile R-valued subgroups of M need not be a facile R-valued subgroup of M, as the following example, due to [85], shows: let $M = (\mathbb{Z}/(2) \times \mathbb{Z}/(2), +)$ be the Klein Four-Group and let $R = (\mathbb{I}, \vee, \wedge)$. Pick elements $t_0 > t_1 > \cdots > t_5$ in $\mathbb{I}$ and define $f, g \in R^M$ as follows:

(1) $f(0,0) = t_1$, $f(1,0) = t_3$, $f(0,1) = f(1,1) = t_4$;
(2) $g(0,0) = t_0$, $g(0,1) = t_2$, $g(1,0) = g(1,1) = t_5$.

Then f and g are $\mathbb{I}$-valued facile subgroups of M. On the other hand,

(3) $(f \vee g)(0,0) = t_0$, $(f \vee g)(1,0) = t_3$, $(f \vee g)(0,1) = t_2$, and $(f \vee g)(1,1) = t_4$

so $f \vee g$ is not an $\mathbb{I}$-valued facile subgroup of M. The general question of when an $\mathbb{I}$-valued subgroup f of a group M can be written as $f_1 \vee f_2$ for $\mathbb{I}$-valued subgroups f_1 and f_2 of M is discussed in [354].

(6.10) PROPOSITION. *Let $(M, *)$ be a group and let R be a QLO-semiring. If U is a directed set of R-valued subgroups of M then $\sum U$ is also an R-valued subgroup of M.*

PROOF. The proof is essentially the same as that of Proposition 5.6. □

Let R be a semiring and let $(M, *)$ be a group such that (R, M) is a convolution context. If $f \in R^M$, let $\hat{g} \in R^M$ be the function defined by $g: m \mapsto f(m) + f(m^{-1})$. Then $g(m) = g(m^{-1})$ for all $m \in M$ and $f \preceq g$. Moreover, $\hat{g}$ is an R-valued subsemigroup of M also having this property and so is an R-valued subgroup of M. If g' is an R-valued subgroup of M satisfying $f \preceq g'$ then it is easily checked that $\hat{g} \preceq g'$. Thus $\hat{g}$ is the R-valued subgroup of M *generated* by f.

For example, let $(M, *)$ be a group with identity element e and let $R = (\mathbb{I}, \vee, \wedge)$. Let $f \in R^M$ and, for each $h \in \mathbb{I}$, let M_h be the subgroup of M generated by $L_h(f)$. Then the $\mathbb{I}$-valued subgroup of M generated by f is given by

$$m \mapsto \begin{cases} 1 & \text{if } m = e \\ \vee\{h \mid m \in M_h\} & \text{otherwise} \end{cases}.$$

See [222] for details.

(6.11) PROPOSITION. *Let $(R, +, \cdot)$ be a semiring and let $(M, *)$ be a group. If $f, g \in R^M$ are [facile, regular] R-valued subgroups of M, one of which is central, then fg is an [facile, regular] R-valued subgroup of M.*

PROOF. This is a direct consequence of Proposition 5.5. □

Let R be a semiring and let $(M, *)$ be a group. An R-valued subgroup f of M is *reducible* if and only if there exist R-valued subgroups g_1 and g_2 of M, neither of which equals f, such that $f = g_1 g_2$. Otherwise it is *irreducible*. The following example is a generalization of one given in [204]:

Let $t_0 \succ t_1$ be central idempotent elements of a semiring R and let D be the dihedral group generated by the pair $\{d_1, d_2\}$ of elements satisfying the relations $d_1^3 = e = d_2^2$ and $d_2 d_1 = d_1^2 d_2$, where e is the identity element of D. Let f be the facile R-valued subgroup of D defined by

$$f: d \mapsto \begin{cases} t_0 & \text{if } d = e \\ t_1 & \text{otherwise} \end{cases}.$$

Then $f = g_1 g_2$, where

$$g_h: d \mapsto \begin{cases} t_0 & \text{if } d \text{ is in the subgroup of } D \text{ generated by } d_h \\ t_1 & \text{otherwise} \end{cases}$$

for $h = 1, 2$.

If $1 \succ t_0$ then each g_h is also reducible. For example, $g_1 = h_1 h_2$, where $h_1: d \mapsto t_0$ for all $d \in D$ and

$$h_2: d \mapsto \begin{cases} 1_R & \text{if } d \text{ is in the subgroup of } D \text{ generated by } d_1 \\ t_1 & \text{otherwise} \end{cases}.$$

(6.12) PROPOSITION. *Let R be a semiring, let $(M, *)$ be a group with identity element e, and let $f \in R^M$ be a central R-valued subgroup of M the image of which contains at least three idempotents: $f(e) = t_0 \succ t_1 \succ t_2$. Then f is reducible.*

PROOF. Under the given hypothesis, it is easy to verify that $f = g_1 g_2$, where

$$g_1: m \mapsto \begin{cases} t_0 & \text{if } f(m) \succeq t_1 \\ f(m) & \text{otherwise} \end{cases}$$

and

$$g_2: m \mapsto \begin{cases} f(m) & \text{if } f(m) \succeq t_1 \\ t_2 & \text{otherwise} \end{cases}.$$

$\square$

If R is a semiring and $(M, *)$ is a group satisfying the condition that (R, M) is a convolution context then we note that

$$(f \langle * \rangle g)(m) = \sum_{x \in M} f(x) g(x^{-1} * m)$$

for each $m \in M$. In particular, if $f, g \in R^M$ are R-valued facile subgroups of M, one of which is central, then

$$(f \langle * \rangle g)(m) = \sum_{x \in M} f(x) g(x^{-1} * m)$$
$$= \sum_{x \in M} f(x^{-1}) g(m^{-1} * x)$$
$$= \sum_{x \in M} g(m^{-1} * x) f(x^{-1})$$
$$= (g \langle * \rangle f)(m^{-1})$$

for all $m \in M$. Thus, in particular, we see that if $f\langle *\rangle g$ is an R-valued facile subgroup of M then $(g\langle *\rangle f)(m^{-1}) = (g\langle *\rangle f)(m)$ for each $m \in M$ and so $f\langle *\rangle g = g\langle *\rangle f$.

The following result is based on [262].

(6.13) PROPOSITION. *Let R be a lattice-ordered semiring and let $(M, *)$ be a group satisfying the condition that (R, M) is a convolution context. Let $f, g \in R^M$ be facile R-valued subgroups of M. Then $f\langle *\rangle g$ is an R-valued facile subgroup of M if and only if $f\langle *\rangle g = g\langle *\rangle f$.*

PROOF. We have already noted above that if $f\langle *\rangle g$ is a facile R-valued subgroup of M then $f\langle *\rangle g = g\langle *\rangle f$. Conversely, assume that $f\langle *\rangle g = g\langle *\rangle f$. By the corollary to Proposition 5.8, we know that $f\langle *\rangle g$ is an R-valued submonoid of M. Also, by the above discussion, $(f\langle *\rangle g)(m) = (g\langle *\rangle f)(m^{-1}) = (f\langle *\rangle g)(m^{-1})$ for each $m \in M$. Finally, since R is lattice-ordered we have $(f\langle *\rangle g)(e) \succeq f(e)g(e) \succeq f(m_1)g(m_2)$ for all $m_1, m_2 \in M$ and so $(f\langle *\rangle g)(e) \succeq (f\langle *\rangle g)(m)$ for all $m \in M$. Thus $f\langle *\rangle g$ is an R-valued facile subgroup of M. $\square$

In particular, we note that if R is a commutative lattice-ordered semiring and if $(M, *)$ is a group satisfying the condition that (R, M) is a convolution context then $f\langle *\rangle g$ is a facile R-valued subgroup of M for all facile R-valued subgroups f and g of M. If both f and g are regular then so is $f\langle *\rangle g$.

If f and g are R-valued subgroups of a group $(M, *)$ then $f\langle *\rangle g$ need not be, as the following example due to [260], shows: let $M = S_3$ and set $H = \{\iota, \langle 12\rangle\}$, $K = \{\iota, \langle 23\rangle\}$, where ι is the identity of M. Let $\chi_H, \chi_K \in \mathbb{I}^M$ be the characteristic functions on H and K respectively. Since H and K are subgroups of M, these are both regular $\mathbb{I}$-valued subgroups of M. However, if $f = \chi_H\langle \cdot\rangle\chi_K$ then

$$f(\langle 13\rangle) = 0 < 1 = f(\langle 132\rangle)f(\langle 12\rangle)$$

while $\langle 13\rangle = \langle 132\rangle\langle 12\rangle$, and so f is not an $\mathbb{I}$-valued subgroup of M.

(6.14) PROPOSITION. *Let R be a commutative lattice-ordered multiplicatively-idempotent semiring and let $(M, *)$ be a group satisfying the condition that (R, M) is a convolution context. Let G be the set of all facile R-valued subgroups of M. Then $(G, \langle *\rangle)$ is a left R-semimodule.*

PROOF. By Proposition 6.13 we know that G is closed under $\langle *\rangle$ and therefore $(G, \langle *\rangle)$ is an abelian semigroup. If $g_1 \in G$ is defined by

$$g_1 = \chi_{\{e\}}: m \mapsto \begin{cases} 1 & \text{if } m = e \\ 0 & \text{otherwise} \end{cases}.$$

then for any $g \in G$ and any $m \in M$ we have

$$(g\langle *\rangle g_1)(m) = \sum_{x \in M} g(x)g_1(x^{-1}m) = g(m)$$

and so $(G, \langle * \rangle)$ is an abelian monoid. Finally, if $r \in R$ and $f, g \in G$ then for all $m \in M$ we have

$$(rf\langle *\rangle rg)(m) = \sum_{m'*m''=m} (rf)(m')(rg)(m'')$$

$$= r^2 \left[\sum_{m'*m''=m} f(m')g(m'') \right]$$

$$= [r^2(f\langle *\rangle g)](m)$$

and so, since R is multiplicatively-idempotent, we have $rf\langle *\rangle rg = r(f\langle *\rangle g)$. $\square$

Let R be a semiring and let $(M, *)$ be a group. If $f \in R^M$ we define $\bar{f} \in R^M$ by setting $\bar{f}: m \mapsto f(m^{-1})$. If f is a central regular R-valued subgroup of M then for all $m_1, m_2 \in M$ we have

$$\bar{f}(m_1 * m_2^{-1}) = f(m_2 * m_1^{-1}) \succeq f(m_2)f(m_1^{-1}) = f(m_1^{-1})f(m_2) = \bar{f}(m_1)\bar{f}(m_2^{-1})$$

and so $\bar{f}$ is again an R-valued subgroup of M. If $f, g \in R^M$ and if (R, M) is a convolution context then for each $m \in M$ we have

$$\overline{f\langle *\rangle g}(m) = (f\langle *\rangle g)(m^{-1})$$

$$= \sum_{y \in M} f(y)g(y^{-1} * m^{-1})$$

$$= \sum_{y \in M} \bar{g}(m * y)\bar{f}(y^{-1})$$

$$= \sum_{x \in M} \bar{g}(x)\bar{f}(x^{-1} * m)$$

$$= (\bar{g}\langle *\rangle \bar{f})(m)$$

and so $\overline{f\langle *\rangle g} = \bar{g}\langle *\rangle \bar{f}$.

We also note that if R is a semiring, if $(M, *)$ is a group, and if $f \in R^M$ then $\bar{f} \preceq f$ implies $f \preceq \bar{f}$. Indeed, if $\bar{f} \preceq f$ then for each $m \in M$ we have $f(m) = \bar{f}(m^{-1}) \preceq f(m^{-1}) = \bar{f}(m)$. Thus, if R is difference ordered we see that $\bar{f} \preceq f$ implies $\bar{f} = f$.

The following result is based on [105].

(6.15) PROPOSITION. Let R be a commutative QLO-semiring and let $(M, *)$ be a group with identity element e. If $f \in R^M$ is a regular R-valued subgroup of M then the set $G(f)$ of all regular R-valued subgroups g of M satisfying $g \preceq f$ is a QLO-semiring under the operations of $+$ and $\langle * \rangle$.

PROOF. By the remark preceding this proposition, we note that if $g, g' \in G(f)$ then $g\langle * \rangle g'$ is a regular R-valued subgroup of M. Moreover, if $m \in M$ then

$$(g\langle * \rangle g')(m) = \sum_{x \in M} g(x) g'(x^{-1} * m)$$
$$\preceq \sum_{x \in M} f(x) f(x^{-1} * m)$$
$$\preceq \sum_{x \in M} f(x * x^{-1} * m)$$
$$= \sum_{x \in M} f(m)$$
$$= f(m)$$

and so $g\langle * \rangle g' \in G(f)$. Moreover, if $\{g_i \mid i \in \Omega\}$ is a family of elements of $G(f)$ and $m \in M$ then $[\sum_{i \in \Omega} g_i](m) = \sum_{i \in \Omega} g_i(m) \preceq \sum_{i \in \Omega} f(m) = f(m)$ and so $\sum_{i \in \Omega} g_i \in G(f)$.

If $g_1 \in G(f)$ is defined by

$$g_1 : m \mapsto \begin{cases} 1 & \text{if } m = e \\ 0 & \text{otherwise} \end{cases}.$$

then for any $g \in G(f)$ and any $m \in M$ we have

$$(g\langle * \rangle g_1)(m) = \sum_{x \in M} g(x) g_1(x^{-1} m) = g(m)$$

and so $g\langle * \rangle g_1 = g$. Similarly, $g_1 \langle * \rangle g = g$.

Finally, if $g \in G(f)$ and if $\{g_i \mid i \in \Omega\} \subseteq G(f)$ then for each $m \in M$ we have

$$\left[g\langle * \rangle \left(\sum_{i \in \Omega} g_i \right) \right](m) = \sum_{x \in M} \left[g(x) \left(\sum_{i \in \Omega} g_i(x^{-1} * m) \right) \right]$$
$$= \sum_{x \in M} \sum_{i \in \Omega} g(x) g_i(x^{-1} * m)$$
$$= \sum_{i \in \Omega} \sum_{x \in M} g(x) g_i(x^{-1} * m)$$
$$= \left[\sum_{i \in \Omega} (g\langle * \rangle g_i) \right](m)$$

which suffices to prove the desired result. $\square$

(6.16) PROPOSITION. Let R be a semiring and let $(M, *)$ be a group satisfying the condition that (R, M) is a convolution context. If $f \in R^M$ is an R-valued regular subgroup of M then $f\langle *\rangle f = f$.

PROOF. By Proposition 5.8, we know that $f\langle *\rangle f \preceq f$ and so we are left to prove the reverse inequality. By Proposition 6.13 we know that $f\langle *\rangle f$ is an R-valued facile subgroup of M. Thus, if $m \in M$, we have

$$(f\langle *\rangle f)(m) = \sum_{x \in M} f(x)f(x^{-1} * m) \succeq f(m)f(e) = f(m)$$

and so $f\langle *\rangle f \succeq f$, as desired. □

Actually, we have proven somewhat more.

(6.17) PROPOSITION. Let R be a semiring and let $(M, *)$ be a group satisfying the condition that (R, M) is a convolution context. If $f, g \in R^M$ are R-valued regular subgroups of M then $f\langle *\rangle g \succeq f$ and $f\langle *\rangle g \succeq g$. Moreover, if $h \in R^M$ is an R-valued regular semigroup of M satisfying $h \succeq f$ and $h \succeq g$ then $h \succeq f\langle *\rangle g$.

PROOF. If $m \in M$ then

$$(f\langle *\rangle g)(m) = \sum_{x \in M} f(x)g(x^{-1} * m) \succeq f(m)g(e) = f(m)$$

so $f\langle *\rangle g \succeq f$. Similarly $f\langle *\rangle g \succeq g$.

If $h \in R^M$ is an R-valued regular semigroup of M satisfying $h \succeq f$ and $h \succeq g$ then for each $m \in M$ we have, by Proposition 6.16,

$$(f\langle *\rangle g)(m) = \sum_{x \in G} f(x)g(x^{-1} * m) \preceq \sum_{x \in G} h(x)h(x^{-1} * m) = (h\langle *\rangle h)(m) = h(m)$$

so $f\langle *\rangle g \preceq h$. □

The following is a generalization of a result of [4].

(6.18) PROPOSITION. Let R be a QLO-semiring and let $(M, *)$ be a group. If $f, g, h \in R^M$ are facile R-valued subgroups of M satisfying $g \preceq fg$ and such that f is central, then $f(g\langle *\rangle h) = g\langle *\rangle(fh)$.

PROOF. If $m \in M$ then

$$[f(g\langle * \rangle h)](m) = f(m) \sum_{x \in M} g(x)h(x^{-1} * m)$$

$$= \sum_{x \in M} f(m)g(x)h(x^{-1} * m)$$

$$\succeq \sum_{x \in M} f(x)f(x^{-1} * m)g(x)h(x^{-1} * m)$$

$$\succeq \sum_{x \in M} f(x)g(x)f(x^{-1} * m)h(x^{-1} * m)$$

$$\succeq \sum_{x \in M} g(x)(fh)(x^{-1} * m)$$

$$= (g\langle * \rangle fh)(m).$$

On the other hand,

$$[g\langle * \rangle(fh)](m) = \sum_{x \in M} g(x)(fh)(x^{-1} * m)$$

$$= \sum_{x \in M} g(x)f(x^{-1} * m)h(x^{-1} * m)$$

$$\succeq \sum_{x \in M} g(x)h(x^{-1} * m)f(x^{-1})f(m)$$

$$= f(m)\left[\sum_{x \in M} g(x)h(x^{-1} * m)f(x^{-1})\right]$$

$$= f(m)\left[\sum_{x \in M} g(x)h(x^{-1} * m)f(x)\right]$$

$$= f(m)\left[\sum_{x \in M} g(x)f(x)h(x^{-1} * m)\right]$$

$$\succeq f(m)\left[\sum_{x \in M} g(x)h(x^{-1} * m)\right]$$

$$= [f(g\langle * \rangle h)](m)$$

and so we have equality. □

Let $(M, *)$ and $(N, \diamond)$ be groups. Then $(M \times N, \diamond)$ is also a group, where the operation $\diamond$ is defined by $(m, n) \diamond (m', n') = (m * m', n \diamond n')$. If R is a semiring such that $(R, M \times N)$ is a convolution context, and if $f, g \in R^{M \times N}$ are R-valued relations between M and N, we can then define the R-valued relation $f\langle \diamond \rangle g \in R^{M \times N}$ by setting

$$(f\langle \diamond \rangle g) : (m, n) \mapsto \sum_{(t,x) \in M \times N} f(t, x)g(t^{-1} * m, x^{-1} \diamond n).$$

Thus, if R is a CLO-semiring and if $(M, *)$ is a group, then we see that $f \geq f\langle \diamond \rangle f$ if and only if $f(m, m') \geq f(t, x)f(t^{-1} * m, x^{-1} * m')$ for all $m, m', t, x \in M$.

(6.19) Proposition. Let R be a QLO-semiring and let N be a normal subgroup of a group $(M, *)$. If $f \in R^M$ is an R-valued facile subgroup of M then the function $g \in R^{M/N}$ defined by $g: m * N \mapsto \sum_{n \in N} f(m * n)$ is a facile R-valued subgroup of M/N.

PROOF. It is easy to verify that the function g is indeed well-defined. If $m_1, m_2 \in M$ then

$$f(m_1 * m_2) = \sum_{n \in N} g(m_1 * m_2 * n)$$
$$= \sum_{n_1 \in N} \sum_{n_2 \in N} g(m_1 * m_2 * n_1 * n_2)$$
$$= \sum_{n_1 \in N} \sum_{n_2 \in N} g(m_1 * n_1 * m_2 * n_2)$$
$$\succeq \sum_{n_1 \in N} \sum_{n_2 \in N} g(m_1 * n_1) g(m_2 * n_2)$$
$$= \left[\sum_{n_1 \in N} g(m_1 * n_1) \right] \left[\sum_{n_2 \in N} g(m_2 * n_2) \right]$$
$$= f(m_1) f(m_2).$$

Moreover, if $m \in M$ then

$$f(m^{-1}) = \sum_{n \in N} g(m^{-1} * n) = \sum_{n \in N} g(n * m^{-1})$$
$$= \sum_{n \in N} g(n^{-1} * m^{-1}) = \sum_{n \in N} g(m * n) = f(m).$$

Finally, if $m \in M$ then $f(m) = \sum_{n \in N} g(m * n) \preceq \sum_{n \in N} g(e) = g(e)$. $\square$

This construction is used by Garzon and Muganda [118] to construction presentations of $\mathbb{I}$-valued facile subgroups of a group.

Normal R-valued subgroups. Let R be a semiring and let $(M, *)$ be a group with identity element e. An R-valued subgroup f of M defines a relation $\sim_f$ on M by setting $m \sim_f m'$ if and only if $f(x * m * y) = f(x * m' * y)$ for all $x, y \in M$. The following result is based on [425].

(6.20) Proposition. Let R be a semiring and let $(M, *)$ be a group with identity element e. If $f \in R^M$ is an R-valued subgroup of M then $N = \{m \in M \mid m \sim_f e\}$ is a normal subgroup of M.

PROOF. Let $n, n' \in N$. Then $f(x*n*y) = f(x*y) = f(x*n'*y)$ for all $x, y \in M$ and so $f(x*n*n'*y) = f((x*n)*n'*y) = f(x*n*y) = f(x*y)$ for all $x, y \in M$ and so

$n*n' \in N$. Moreover, $f(x*n^{-1}*y) = f(x*(n^{-1}*y)) = f(x*n*(n^{-1}*y)) = f(x*y)$ for all $x, y \in M$ and so $n^{-1} \in N$. Thus N is a subgroup of M. If $n \in N$ and if $z \in M$ then

$$f(x*(z^{-1}*n*z)*y) = f((x*z^{-1})*n*(z*y)) = f(x*z^{-1}*z*y) = f(x*y)$$

for all $x, y \in M$ and so $z^{-1}*n*z \in N$, showing that N is normal in M. □

It is easily seen that, in the situation of Proposition 6.13, we have $m \sim_f m'$ in M if and only if $m*N = m'*N$.

Let $(M, *)$ be a group and let R be a semiring. A facile R-valued subgroup $f \in R^M$ is *normal* if and only if $f(m_1^{-1}*m_2*m_1) = f(m_2)$ for all $m_1, m_2 \in M$. That is to say, f is normal if and only if it coincides with each of its conjugates. If M is abelian then, of course, every facile R-valued subgroup of M is normal.

(6.21) PROPOSITION. *If $(M, *)$ is a group and if R is a semiring then a facile R-valued subgroup $f \in R^M$ is normal if and only if $f(m_1*m_2) = f(m_2*m_1)$ for all $m_1, m_2 \in M$.*

PROOF. If f is normal and $m_1, m_2 \in M$ then

$$f(m_1*m_2) = f(m_1^{-1}*(m_1*m_2)*m_1) = f(m_2*m_1).$$

Conversely, if the given condition is satisfied and $m_1, m_2 \in M$ then

$$f(m_1^{-1}*m_2*m_1) = f((m_2*m_1)*m_1^{-1}) = f(m_2),$$

as desired. □

The following example is in [332]: let $M = S_4$ and define $f \in \mathbb{I}^M$ as follows:

$$f(\sigma) = \begin{cases} 1 & \text{if } \sigma \text{ is the identity map} \\ \frac{1}{2} & \text{if } \sigma = \langle 13 \rangle \\ \frac{1}{3} & \text{if } \sigma \in \{\langle 24 \rangle, \langle 13 \rangle \langle 24 \rangle\} \\ \frac{1}{4} & \text{if } \sigma \in \{\langle 12 \rangle \langle 34 \rangle, \langle 14 \rangle \langle 23 \rangle, \langle 1234 \rangle, \langle 1432 \rangle\} \\ 0 & \text{otherwise} \end{cases}.$$

Then f is an $\mathbb{I}$-valued subgroup of M but which is not normal since

$$f((\langle 24 \rangle \langle 34 \rangle)\langle 1234 \rangle) = f(\langle 13 \rangle) = \frac{1}{2} \neq \frac{1}{3} = f(\langle 24 \rangle) = f(\langle 1234 \rangle(\langle 24 \rangle \langle 34 \rangle)).$$

Also, consider the following example, due to [202]: let $M = D_4$, the dihedral group with generators $\{a, b\}$ subject to the relations $a^4 = e = b^2$ and $b*a = a^3*b$. Select $t_0 > t_1$ in $\mathbb{I}$ and let $f \in \mathbb{I}^M$ be the function defined by

$$f: m \mapsto \begin{cases} t_0 & \text{if } m \text{ is in the subgroup of } M \text{ generated by } b \\ t_1 & \text{otherwise} \end{cases}.$$

Then $f(a * (a * b)) = f(a^2 * b) = t_1 \neq t_0 = f(b) = f((a * b) * a)$ and so f is not normal.

Note that if $(M, *)$ is a group, if R is a semiring, if f is an R-valued subgroup of M, and if r is a multiplicatively-idempotent element of R satisfying the condition that $L_r(f) \neq \emptyset$, then f is normal if and only if $L_r(f)$ is a normal subgroup of M. This observation has several implications. For example, if M is a simple group then a normal R-valued subgroup of M must be constant everywhere except possibly at the identity element of M. For other such results, refer to [220].

(6.22) PROPOSITION. *Let R be a semiring and let $(M, *)$ be a group satisfying the condition that (R, M) is a convolution context. Let $f, g \in R^M$ be facile R-valued subgroups of M satisfying the condition that g is normal and one of them is central. Then $f \langle * \rangle g$ is an R-valued normal facile subgroup of M.*

PROOF. Since one of the functions f and g is central, we have $f \langle * \rangle g = g \langle * \rangle f$. Hence, by Proposition 6.13, we know that $f \langle * \rangle g$ is an R-valued facile subgroup of M, and so all we have to establish is normality. Indeed, if $m_1, m_2 \in M$ then

$$\begin{aligned}(f \langle * \rangle g)(m_1 * m_2) &= \sum_{m \in M} f(m) g(m^{-1} * m_1 * m_2) \\ &= \sum_{m \in M} f(m) g(m_2 * m_1 * m^{-1}) \\ &= \sum_{m \in M} g(m_2 * m_1 * m^{-1}) f(m) \\ &= (g \langle * \rangle f)(m_2 * m_1) \\ &= (f \langle * \rangle g)(m_2 * m_1)\end{aligned}$$

which proves the result. □

Putting Propositions 6.22 and 6.16 together, we see that if R is a commutative semiring and $(M, *)$ is a group satisfying the condition that (R, M) is a convolution context then the set of all normal facile R-valued subgroups of M is a commutative semigroup under the operation $\langle * \rangle$. Moreover, in this semigroup all regular elements are idempotent.

Let M be a finite group and let f be a regular $\mathbb{I}$-valued subgroup of M. For each $t \in \mathbb{I}$ let the subgroup $L_t(f) = \{m \in M \mid f(m) \geq t\}$. If f is normal then, following [12], we note that:

(1) The number of distinct subgroups of M of the form $L_t(f)$ is finite, say

$$L_{t_1}(f) \subset L_{t_2}(f) \subset \cdots \subset L_{t_k}(f).$$

(2) For each $1 \leq h < k$, we have $L_{t_h}(f) \triangleleft L_{t_{h+1}}$.

The converse of this is false. To see this, let M be the alternating group A_4 with identity element ι and define $f \in \mathbb{I}^M$ by setting:

$f(\iota) = 1$;
$f((12)(34)) = \frac{1}{2}$;
$f((14)(23)) = f((13)(24)) = \frac{1}{3}$;
$f(ijk) = 0$ for all distinct $i, j, k \in \{1, 2, 3, 4\}$.

Then conditions (1) and (2) are satisfied but f is not a normal $\mathbb{I}$-valued subgroup of M.

(6.23) PROPOSITION. Let $(M, *)$ be a group and let R be a semiring. Let $f \in R^M$ be an facile normal R-valued subgroup of M and let $r \in R$ be an idempotent element satisfying the condition that $L_r(f)$ is nonempty. Then the relation $\equiv$ on M defined by $m_1 \equiv m_2$ if and only if $m_1 * m_2^{-1} \in L_r(f)$ is a congruence relation on M.

PROOF. If $m \in M$ then $f(m * m^{-1}) = f(e) \in L_r(f)$ since $L_r(f) \neq \emptyset$ and so $m \equiv m$. If $m_1 \equiv m_2$ then $r \preceq f(m_1 * m_2^{-1}) = f((m_1 * m_2^{-1})^{-1}) = f(m_2 * m_1^{-1})$ and so $m_2 \equiv m_2$. If $m_1 \equiv m_2$ and $m_2 \equiv m_3$ then

$$f(m_1 * m_3^{-1}) = f((m_1 * m_2^{-1}) * (m_2 * m_3^{-1})) f(m_1 * m_2^{-1}) f(m_2 * m_3^{-1}) \succeq r^2 = r$$

so $m_1 \equiv m_3$. Thus $\equiv$ is an equivalence relation on M.

If $m_1 \equiv m_2$ and if $m \in M$ then $f((m_1 * m) * (m_2 * m)^{-1}) = f(m_1 * m_2^{-1}) \succeq r$ and so $m_1 * m \equiv m_2 * m$. Similarly, $m * m_1 \equiv m * m_2$. Thus $\equiv$ is a congruence relation. $\square$

(6.24) PROPOSITION. Let R be a semiring and let $(M, *)$ be a group satisfying the condition that (R, M) is a convolution context. Let $f, g \in R^M$ be facile normal R-valued subgroups of M, one of which is central. If $f\langle*\rangle g$ is an R-valued facile subgroup of M then it is normal.

PROOF. Let $m_1, m_2 \in M$. If $x, y \in M$ then $x * y = m_2$ if and only if

$$(m_1^{-1} * x * m_1) * (m_1^{-1} * y * m_1) = m_1^{-1} * m_2 * m_1$$

and so

$$(f\langle*\rangle g)(m_2) = \sum_{x*y=m_2} f(x)g(y)$$

$$= \sum_{x*y=m_2} f(m_1^{-1} * x * m_1) g(m_1^{-1} * y * m_1)$$

$$\preceq \sum_{s*t=m_1^{-1}*m_2*m_1} f(s)g(t)$$

$$= (f\langle*\rangle g)(m_1^{-1} * m_2 * m_1).$$

Conversely, if $s, t \in M$ then $s, t = m_1^{-1} * m_2 * m_1$ if and only if

$$(m_1 * s * m_1^{-1}) * (m_1 * t * m_1^{-1}) = m_2$$

and so a similar argument shows that

$$(f\langle*\rangle g)(m_1^{-1} * m_2 * m_1) \preceq (f\langle*\rangle g)(m_2),$$

establishing equality. Therefore $f\langle*\rangle g$ is normal. □

The following example is due to [331]: let $M = (\mathbb{R}^{[-1,1]}, +)$ and for each $r \in [0, 1]$ let $M_r = \{\psi \in M \mid \psi(b) = 0 \text{ for all } -r \leq b \leq r\}$. Then surely M_r is a subgroup of M. Define the $\mathbb{I}$-valued facile subgroup f of M as follows:

$$f: \psi \mapsto \begin{cases} 1 & \text{if } \psi \text{ is the 0-function} \\ \frac{1}{n+1} & \text{if } \psi \in M_{1/(n+1)} \setminus M_{1/n} \\ 0 & \text{otherwise.} \end{cases}$$

For each $0 < r \leq 1$, we note that $f^{-1}([r, 1])$ is a proper subgroup of M.

(6.25) PROPOSITION. [331] Let R be an additively-idempotent semiring, let $(M, *)$ and $(N, \diamond)$ be groups, and let $u: M \to N$ be a group epimorphism with kernel K. If $f \in R^M$ is a facile normal R-valued subgroup of M then $h_u[f]$ is a facile normal R-valued subgroup of N.

PROOF. Set $g = h_u[f]$. If $n = u(m)$ is an element of N then

$$g(n) = \sum_{u(m)=n} f(m) = \sum_{k \in K} f(m * k).$$

If $m_1, m_2 \in M$ and if $n_i = u(m_i)$ for $i = 1, 2$. Then

$$g(n_1 \diamond n_2) = g(u(m_1 * m_2))$$
$$= \sum_{k \in K} f(m_1 * m_2 * k)$$
$$= \sum_{k \in K} f(m_2 * k * m_1)$$
$$= \sum_{k \in K} f(m_2 * m_1 * k)$$
$$= g(u(m_2 * m_1))$$
$$= g(n_2 \diamond n_1)$$

and so g is a normal R-valued subgroup of N. □

Let R be a semiring and let $(M, *)$ and $(N, \diamond)$ be groups. If f and g are normal R-valued subgroups of M and N respectively, one of which is central, then we know that $f \times g \in R^{M \times N}$ is an R-valued subgroup of $M \times N$, which is also clearly normal.

We now consider a construction originally presented in [335]. Let R be a semiring and let $(M, *)$ be a group. If $g \preceq f$ are R-valued subgroups of M then we say that g is *normal in* f, and write $g \triangleleft f$, if and only if $g(m' * m'') = g(m'' * m')$ for all $m', m'' \in supp(f)$. By Proposition 6.21, we see that an R-valued subgroup g of M is normal if and only if $g \triangleleft f_1$, where f_1 is the constant function $f_1: m \mapsto 1$.

Now assume that R is a CLO-semiring (if M is finite then completeness is not necesary) and let $g \triangleleft f$ be R-valued subgroups of M. The semiring R is both entire and zerosumfree so, by Proposition 6.2, we know that $supp(g) \subseteq supp(f)$ are subgroups of M. Moreover, $supp(g)$ is normal in $supp(f)$ for if $x \in supp(f)$ and $m \in supp(g)$ then $x * m * x^{-1} \in supp(f)$. This implies that

$$g(x * m * x^{-1}) = g(x^{-1} * x * m) = g(m) \neq 0$$

and so $x * m * x^{-1} \in supp(g)$. Let $\mathcal{C}$ be the set of all subgroups N of M containing $supp(f)$ as a subgroup and $supp(g)$ as a normal subgroup. This set is nonempty since $supp(f) \in \mathcal{C}$. If N_0 is the subgroup of M generated by $\cup \{N \mid N \in \mathcal{C}\}$ then $N_0 \in \mathcal{C}$ and so N_0 is clearly the unique maximal element of $\mathcal{C}$, which we will denote by $N(g, f)$.

Let $H = supp(g)$ and define a function $q: N(g, f)/H \to R$ by

$$q: x * H \mapsto \sum_{m \in H} f(x * m).$$

If $x, y \in N(g, f)$ and $m, m' \in H$ then

$$f(x * m)f(y * m') \preceq f(x * m * y * m')$$
$$= f(x * y * y^{-1} * m * y * m')$$
$$\preceq \sum_{t \in H} x * y * t$$
$$= q(x * y * H)$$

and so

$$q(x * H)q(y * H) = \sum_{m \in H} \sum_{m' \in H} f(x * m)f(y * m') \preceq q(x * y * H).$$

This shows that q is an R-valued subsemigroup of $N(g,f)/H$. Moreover, if $x \in N(g,f)$ then

$$q(x^{-1} * H) = \sum_{m \in H} f(x^{-1} * m)$$
$$= \sum_{m \in H} f(m^{-1} * x)$$
$$= \sum_{m \in H} f(x * x^{-1} * m^{-1} * x)$$
$$= \sum_{m \in H} f(x * m)$$
$$= q(x * H).$$

Therefore q is an R-valued subgroup of $N(g,f)/H$, which we will denote by f/g. Clearly $supp(f/g) = supp(f)/supp(g)$. Also, if $f = f_1$ (namely the constant function $m \mapsto 1$) then f/g is just the constant function on M/H.

An R-valued subsemigroup f of a group $(M, *)$ if *self-normal* if and only if $f \triangleleft f$, i.e., if and only if $f(m * m') = f(m' * m)$ for all $m, m' \in supp(f)$. Thus, if $R = \mathbb{B}$ then every R-valued subsemigroup f of M is self-normal.

Let $R = (\mathbb{R}^+, +, \cdot)$ and let k be a positive integer. Let M be the set of all $k \times k$ matrices A over $\mathbb{R}$ having determinant 1. Then M is a group under ordinary matrix multiplication and the function $f: A \mapsto |A|$ is a self-normal R-valued subgroup of M.

The following example, due to [332], shows that if f and g are self-normal, $f\langle*\rangle g$ need not be: let $M = S_4$. Define $f, g \in \mathbb{I}^M$ as follows:

$$f: \sigma \mapsto \begin{cases} 1 & \text{if } \sigma \text{ is the identity map} \\ \frac{1}{2} & \text{if } \sigma = \langle 12\rangle\langle 34\rangle \\ \frac{1}{6} & \text{if } \sigma \in \{\langle 13\rangle\langle 24\rangle, \langle 14\rangle\langle 23\rangle\} \\ 0 & \text{otherwise} \end{cases}.$$

and

$$g: \sigma \mapsto \begin{cases} \frac{1}{3} & \text{if } \sigma \text{ is the identity map} \\ \frac{1}{4} & \text{if } \sigma \in \{\langle 12\rangle\langle 34\rangle, \langle 13\rangle\langle 24\rangle, \langle 14\rangle\langle 23\rangle\} \\ \frac{1}{5} & \text{if } \sigma \text{ is a three-cycle} \\ 0 & \text{otherwise} \end{cases}.$$

Then f and g are both self-normal but $f\langle*\rangle g$ is not.

As an immediate consequence of the definition, we see that if R is a lattice-ordered semiring and if $f, g \in R^M$ are self-normal R-valued subgroups of R then

$f \vee g$ is also self-normal. If R is a CLO-semiring, this does not, howevere extend to infinite meets [332]: let $M = S_4$ and let ι be the identity map in M. Then

$$N = \{\iota, \langle 12\rangle\langle 34\rangle, \langle 13\rangle\langle 24\rangle, \langle 14\rangle\langle 23\rangle, \langle 13\rangle, \langle 24\rangle, \langle 1234\rangle, \langle 1432\rangle\}$$

is a subgroup of M. For each positive integer k define the self-normal $\mathbb{I}$-valued subgroup f_k of N as follows:

$$f_k : \sigma \mapsto \begin{cases} 1 & \text{if } \sigma \in \{\iota, \langle 12\rangle\langle 34\rangle\} \\ \frac{1}{k} & \text{if } \sigma \in \{\langle 13\rangle\langle 24\rangle, \langle 14\rangle\langle 23\rangle\} \\ 0 & \text{otherwise} \end{cases}.$$

Set $f = \bigwedge_{k=1}^{\infty} f_k$. Then $f(\iota) = f(\langle 12\rangle\langle 34\rangle) = 1$ and $f(\sigma) = 0$ otherwise, so f is not self-normal.

Using this definition of normality, Ray [335] has constructed a theory of solvable $\mathbb{I}$-valued subgroups of a group.

Weights. Let $(M, *)$ be a group and let $subg(M)$ be the complete lattice of all subgroups of M, in which, for each family $\{N_i \mid i \in \Omega\}$ of members of $subg(M)$, we have that $\bigwedge_{i \in \Omega} N_i$ is just $\bigcap_{i \in \Omega} N_i$ and $\bigvee_{i \in \Omega} N_i$ is the subgroup of M generated by $\bigcup_{i \in \Omega} N_i$. For each $m \in M$, let $S(m) = \{N \in subg(M) \mid m \in N\}$. Then $S(m) \neq \varnothing$ for each $m \in M$ and $S(m) = subg(M)$ if and only if m is the identity element e of M. Moreover, we have an equivalence relation $\sim$ on M defined by the condition that $m \sim m'$ if and only if $S(m) = S(m')$.

At the beginning of this chapter, we have already noted that if R is a difference-ordered multiplicatively-idempotent semiring and if $f \in R^M$ is a facile R-valued subgroup of M then $m \sim m'$ implies that $f(m) = f(m')$. If $\{x_i \mid i \in \Lambda\}$ is a full set of representatives of the distinct equivalence classes of M with respect to $\sim$, this means that M can be written as a union of disjoint subsets $\bigcup_{i \in \Lambda} S(x_i)$ on each of which f is constant. If N is a subgroup of M and if $A(x_h) \cap N \neq \varnothing$ then there exists an element n of N belonging to $A(x_h)$. Since $n \sim y$ for each $y \in A(x_h)$, this means that every element of $A(x_h)$ belongs to precisely the same subgroups of M that n does, and so, in particular, $A(x_h) \subseteq N$. Thus each subgroup of M is of the form $\bigcup_{i \in \Lambda'} S(x_i)$ for some subset Λ' of Λ.

Let $\chi_i \in R^M$ be the characteristic function on $S(x_i)$, namely the function defined by

$$\chi_i : m \mapsto \begin{cases} 1 & \text{if } m \in S(x_i) \\ 0 & \text{otherwise} \end{cases}.$$

Then each facile R-valued subgroup f of M can be written in the form

$$f = \sum_{i \in \Lambda} f(x_i) \chi_i.$$

Now let R be a CLO-semiring. An R-valued *weight* on a group $(M, *)$ is a function from $subg(M)$ to R satisfying the condition that $w(\bigvee_{i \in \Omega} N_i) = \bigwedge_{i \in \Omega} w(N_i)$ for any subset $\{N_i \mid i \in \Omega\}$ of $subg(M)$. If $\{w_i \mid i \in \Omega\}$ is a family of R-valued weights on M then the function $\bigwedge_{i \in \Omega} w_i$ from $subg(M)$ to R defined by

$$\bigwedge_{i \in \Omega} w_i: N \mapsto \bigwedge_{i \in \Omega} w_i(N)$$

is, by the completeness of R, also an R-valued weight on M. Therefore the collection of all R-valued weights on M is easily seen to be a complete lattice.

Weights on groups with values in $\mathbb{I}$ were first studied in [227]. We note, in particular, that if $N \subseteq N'$ are subgroups of M then $w(N') = w(N \vee N') = w(N) \wedge w(N') \leq w(N)$.

Let $(R, \vee, \wedge)$ be a complete lattice and let $(M, *)$ be a group with identity element e. Let f be a facile R-valued subgroup of M. As above, we write $M = \bigcup_{i \in \Lambda} S(x_i)$. Then $f = \sum_{i \in \Lambda} f(x_i) \chi_i$, where χ_i is the characteristic function on $S(x_i)$. Define a function $w_f: subg(M) \to R$ by setting $w_f: N \mapsto \wedge \{f(x_i) \mid x_i \in N\}$. This function can be extended to a $sub(M)$-gradation $\bar{w}_f$ of R by setting

$$\bar{w}_f: B \mapsto \begin{cases} f(e) & \text{if } B = \varnothing \\ \wedge \{f(x_i) \mid B \cap S(x_i) \neq \varnothing\} & \text{otherwise} \end{cases}.$$

Following [227], we assert that w_f is an R-valued weight on M.

To prove this assertion, our first claim is that $\bar{w}_f(B) = w_f(\langle B \rangle)$ for each subset B of M, where here $\langle B \rangle$ denotes the subgroup of M generated by B. Indeed, this is trivial of $B = \varnothing$ so assume that $B \neq \varnothing$. Since $B \subseteq \langle B \rangle$ we have $w_f(\langle B \rangle) \leq \bar{w}_f(B)$. Assume that this inequality is strict. Then there exists an $h \in \Lambda$ such that $S(x_h) \subseteq \langle B \rangle$ and $\bar{w}_f(B) \not\leq f(x_h)$. Since $x_h \in \langle B \rangle$, there exist elements $b_1, \ldots, b_t$ in B and integers $k_1, \ldots, k_t$ such that $x_h = b_1^{k_1} \cdot \ldots \cdot b_t^{k_t}$ and hence $f(x_h) \geq f(b_1) \wedge \cdots \wedge f(b_t) \geq \bar{w}_f(B)$. This is a contradiction and so we must have $\bar{w}_f(B) = w_f(\langle B \rangle)$, as desired.

Now let $\{N_i \mid i \in \Omega\}$ be a family of subgroups of M. By the above remark, we note that $w_f(\bigvee_{i \in \Omega} N_i) = \bar{w}_f(\bigcup_{i \in \Omega} N_i) \leq w_f(N_t)$ for all $t \in \Omega$. Hence $w_f(\bigvee_{i \in \Omega} N_i) \leq \bigwedge_{i \in \Omega} w((N_i)$. Suppose the inequality is strict. Then there exists an $h \in \Lambda$ such that $S(x_h) \subseteq \bigvee_{i \in \Omega} N_i$ and $f(x_h) \not\geq \bigwedge_{i \in \Omega} w_f(N_i)$. This means that $S(x_h) \cap N_i = \varnothing$ for each $i \in \Omega$ and so $S(x_i) \cap [\bigcup_{i \in \Omega} N_i] = \varnothing$, which is a contradiction. This proves that we have equality and so w_f is an R-valued weight on M, as asserted.

Congruences. We have already considered R-valued congruence relations on semigroups, and now want to do so on groups. For example, let $(M, *)$ be a group and let R be a semiring. If f is a regular R-valued subgroup of M then the function $f^\sharp \in R^{M \times M}$ defined by $f^\sharp \colon (m, m') \mapsto f(m * m'^{-1})$ is an R-valued congruence relation on M.

If (R, M) is a convolution context we have $f^\sharp \circ g^\sharp = [f \langle * \rangle g]^\sharp$. To see this, we note that if $m, m' \in M$ then

$$(f^\sharp \circ g^\sharp)(m, m') = \sum_{x \in G} f^\sharp(m, x) g^\sharp(x, m')$$

$$= \sum_{x \in G} f(m * x^{-1}) g(x * m'^{-1})$$

$$= \sum_{x * y = m * m'^{-1}} f(x) g(y)$$

$$= (f \langle * \rangle g)(m * m'^{-1})$$

$$= [f \langle * \rangle g]^\sharp(m, m').$$

(6.26) PROPOSITION. [232, 245] *Let $(M, *)$ be a group with identity element e and let R be a difference-ordered semiring. If $f \in R^{M \times M}$ is an R-valued congruence relation on M then, for all $x, y, m, m' \in M$,*

(1) $f(m, m') = f(m * y, m' * y) = f(x * m, x * m') = f(x * m * y, x * m' * y)$,
(2) $f(e, m^{-1} * m') = f(m, m') = f(m'^{-1} * m, e)$,
(3) $f(m, m') = f(m^{-1}, m'^{-1})$.

PROOF. (1) Since f is an R-valued congruence relation, we know that

$$f(m * y, m' * y) \succeq f(m, m') = f(m * y * y^{-1}, m' * y * y^{-1}) \succeq f(m * y, m' * y)$$

and so $f(m, m') = f(m * y, m' * y)$. The other inequalities are proven similarly.
(2) This is immediate from (1).
(3) By (2), $f(m, m') = f(e, m^{-1} * m') = f(m'^{-1}, m^{-1}) = f(m^{-1}, m'^{-1})$. □

(6.27) PROPOSITION. *Let $(M, *)$ be a group and let R be a semiring. If $f, g \in R^{M \times M}$ are R-valued congruence relations on M then so is $f \circ g$.*

PROOF. If $m, m', m'' \in M$ then

$$(f \circ g)(m'' * m, m'' * m') = \sum_{x \in M} f(m'' * m, x) g(x, m'' * m')$$

$$= \sum_{x \in M} f(m'' * m, m'' * x) g(m'' * x, m'' * m')$$

$$= \sum_{x \in M} f(m, x) g(x, m')$$

$$= (f \circ g)(m, m')$$

which proves our assertion. □

(6.28) PROPOSITION. [232] *Let $(M, *)$ be a group having identity element e and let R be a semiring. If $f \in R^{M \times M}$ is an R-valued congruence relation on M then then equivalence class f_e of e is a normal R-valued subgroup of M.*

PROOF. First note that $f_e(e) = f(e, e) = 1$. If $m, m' \in M$ then, repeatedly using Proposition 6.20, we obtain $f_e(m * m'^{-1}) = f(e, m * m'^{-1}) = f(m', m) = f(m, m') \succeq f(m, e)f(e, m') = f(e, m)f(e, m'^{-1}) = f_e(m)f_e(m'^{-1})$, which proves that f_e is an R-valued subgroup of M.

Moreover, $f_e(m * m') = f(e, m * m') = f(m^{-1} * e * m, m^{-1} * (m * m') * m) = f(e, m' * m) = f_e(m' * m)$ and so f_e is normal. □

Recall the notion of an R-valued partition on a set. The following results are based on [290].

(6.29) PROPOSITION. *Let R be a simple semiring and let $(M, *)$ be a group with identity e satisfying the condition that (R, M) is a convolution context. Let $P \subseteq R^M$ be an R-valued partition of M such that $(P, \langle * \rangle)$ is a group. Then for all $m, m' \in M$:*

(1) $u_m \langle * \rangle u_{m'} = u_{m*m'}$;
(2) $p_{m,1} \langle * \rangle u_{m'} = u_{m*m'} = u_m \langle * \rangle p_{m',1}$;
(3) $p_{m,1} \langle * \rangle u_e = u_m = u_e \langle * \rangle p_{m,1}$;
(4) $u_e \langle * \rangle u_m = u_m = u_m \langle * \rangle u_e$;
(5) $u_m \langle * \rangle u_{m^{-1}} = u_e = u_{m^{-1}} \langle * \rangle u_m$.

Moreover, the R-valued equivalence relation h_P defined by P on M is a congruence relation.

PROOF. (1) If $m, m' \in M$ then

$$(u_m \langle * \rangle u_{m'})(m * m') = \sum_{x*y=m*m'} u_m(x) u_{m'}(y) \succeq u_m(m) u_{m'}(m') = 1$$

and so $(u_m \langle * \rangle u_{m'})(m * m') = 1$, proving that $u_m \langle * \rangle u_{m'} = u_{m*m'}$.

(2) If $m, m' \in M$ then

$$(p_{m,1} \langle * \rangle u_{m'})(m * m') = \sum_{x*y=m*m'} p_{m,1}(x) u_{m'}(y)$$
$$= \sum_{m*y=m*m'} u_{m'}(y)$$
$$\succeq u_{m'}(m') = 1$$

and so $p_{m,1} \langle * \rangle u_{m'} = u_{m*m'}$. The second equality is proven similarly.

(3) This is a special case of (2).

(4) If $m \in M$ then
$$(u_e\langle *\rangle u_m)(m) = \sum_{x*y=m} u_e(x)u_m(y) \succeq u_e(e)u_m(m) = 1$$
and so $u_e\langle *\rangle u_m = u_m$. The other equality is proven similarly.

(5) This is a direct consequence of (1).

Finally, we note that if $m, m', m'' \in M$ then
$$\begin{aligned} h_P(m*m'', m'*m'') &= u_{m*m''}(m'*m'') \\ &= (u_m\langle *\rangle u_{m''})(m'*m'') \\ &\succeq u_m(m')u_{m''}(m'') \\ &= u_m(m') = h_P(m,m') \end{aligned}$$
and similarly $h_P(m''*m, m''*m') \succeq h_P(m, m')$. □

(6.30) PROPOSITION. *Let R be a simple CLO-semiring and let $(M, *)$ be a group with identity element e. Let h be a congruence relation on M and let P_h be the R-valued partition of M defined by h. Then $(P_h, \langle *\rangle)$ is a group.*

PROOF. If $m, m', m'' \in M$ then
$$\begin{aligned} (u_m\langle *\rangle u_{m'})(m'') &= \sum_{x \in M} u_m(x)u_{m'}(x^{-1}*m'') \\ &= \sum_{x \in M} h(m,x)h(m', x^{-1}*m'') \\ &= \sum_{x \in M} h(m*m', x*m')h(x*m', m'') \\ &\preceq h(m*m', m'') \\ &= h(m, m''*m'^{-1})h(m', m') \\ &= u_m(m''*m'^{-1})u_{m'}(m') \\ &\preceq (u_m\langle *\rangle u_{m'})(m'') \end{aligned}$$
and so, since R is difference-ordered, we have
$$(u_m\langle *\rangle u_{m'})(m'') = h(m*m', m'') = u_{m*m'}(m'')$$
for all $m, m', m'' \in M$, proving that $u_m\langle *\rangle u_{m'} = u_{m*m'}$ for all $m, m' \in M$. This suffices to show that $(P_h, \langle *\rangle)$ is a semigroup. Moreover, it also implies that $u_e\langle *\rangle u_m = u_m = u_m\langle *\rangle u_e$ and $u_m\langle *\rangle u_{m^{-1}} = u_e = u_{m^{-1}}\langle *\rangle u_m$ for all $m \in M$ and so it is in fact a group. □

Recall that If $(M, *)$ is a group with identity element e, if R is a semiring, and if $f \in R^{M \times M}$ is an R-valued equivalence relation, then the set M/f of all elements of f with respect to f is itself a semigroup with respect to the operation $[*]$ defined by $f_m[*]f_{m'} = f_{m*m'}$.

(6.31) PROPOSITION. Let $(M, *)$ be a group and let R be a complete difference-ordered commutative semiring. Then $f \circ g = g \circ f$ for all R-valued congruence relations f and g on M.

PROOF. If $m, m' \in M$ then

$$(g \circ f)(m, m') = \sum_{x \in M} g(m, x) f(x, m')$$
$$= \sum_{y \in M} g(m, m' * y^{-1} * m) f(m' * y^{-1} * m, m')$$
$$= \sum_{y \in M} f(m' * y^{-1} * m, m' * y^{-1} * y) g(y * y^{-1} * m, m' * y^{-1} * m)$$
$$\succeq \sum_{y \in M} f(m, y) g(y, m')$$
$$= (f \circ g)(m, m').$$

Thus $g \circ f \succeq f \circ g$. Similarly $f \circ g \succeq g \circ f$ and so we have equality. □

As a consequence of this result and Proposition 5.22, we see that if $(M, *)$ is a group and if R is a complete simple difference-ordered commutative semiring then the set of all R-valued congruence relations on M is closed under composition and, moreover, it is a join-semilattice.

Now suppose that $(M, *)$ is a group and that $f, g \in R^{M \times M}$ are R-valued equivalence relations on M, at least one of which is central. As in the case of semigroups, it is easy to see that in this case $fg \in R^{M \times M}$ is also an R-valued congruence relation on M. This allows us to extend Proposition 6.31

(6.32) PROPOSITION. Let $(M, *)$ be a group and let R be a complete difference-ordered multiplicatively-idempotent commutative semiring. Then the set of all R-valued congruence relations on M is a lattice.

PROOF. By Proposition 6.31, we see that the set C of all R-valued congruence relations on M is a join-semilattice, the join in which is given by $\circ$. Since R is multiplicatively-idempotent, we know that $(R, \preceq)$ is a meet-semilattice, the meet in which is given by multiplication. Since multiplication and $\preceq$ are defined componentwise, the same is therefore true for $R^{M \times M}$. Since, as noted above, the set C is closed under multiplication and so we see that if $f, g \in C$ then $fg \preceq f, g$ and $h \preceq fg$ whenever $h \preceq f, g$. Thus $(C, \circ, \cdot)$ is a lattice. □

Kim and Bae [205] show that if $R = \mathbb{I}$ then this lattice is, in fact, modular.

7. Semiring-valued Submodules and Subspaces

R-valued submodules of a module. Let R be a semiring. If S is a ring and $(M, +)$ is a left S-module then an R-valued subgroup $f \in R^M$ of $(M, +)$ is an *R-valued S-submodule* of M if and only if it satisfies the additional condition

(1) $f(sm) \succeq f(m)$ for all $s \in S$ and $m \in M$.

See [426, 427]. In particular, this implies that $f(sm) = f(m)$ for each invertible element s of S. Note that this condition also implies that $f(0_M) = f(0_R m) \succeq f(m)$ and $f(m) = f(-m)$ for each $m \in M$ and so every R-valued S-submodule of M is a facile R-valued subgroup of $(M, +)$. If $f(0_M) = 1_R$, then we say that f is a *regular R-valued S-submodule* of M. Thus, if M is a left S-module and if $\emptyset \neq N \subseteq M$ then $\chi_N \in \mathbb{B}^M$ is a regular $\mathbb{B}$-valued S-submodule of M if and only if N is an S-submodule of M.

Again, let us begin by looking at some examples.

(I) EXAMPLE. Let M be a left S-module for some ring S, let R be a semiring, and let $r \in R$ satisfy $r \succeq r^2$. Given $m_0 \in M$, the function $f \in R^M$ defined by

$$f: m \mapsto \begin{cases} r & \text{if } m \in Sm_0 \\ 0 & \text{otherwise.} \end{cases}$$

is an R-valued S-submodule of M. More generally, given any S-submodule N of M, the function $g \in R^M$ defined by

$$g: m \mapsto \begin{cases} r & \text{if } m \in N \\ 0 & \text{otherwise.} \end{cases}$$

is an R-valued S-submodule of M.

(II) EXAMPLE. If $(M,+)$ is a left S-module then $\mathbb{I}$-valued S-submodules of M are studied in [298]. Methods of constructing $\mathbb{I}$-valued S-submodules of M using (hereditary) torsion theories on the category $S - mod$ of all left S-modules are presented in [143].

(III) EXAMPLE. Let R be a bounded linearly-ordered set with minimal element 0. Then $(R, \wedge, \vee)$ is a semiring. If M is a left S-module for some ring S, then the R-valued S-submodules of M were studied by Fleischer [107] under the name of *normed modules*. Normed modules with values in $\mathbb{I}$ arise naturally in the study of modules with nonnegative filtrations.

Let R be a semiring and let M be a left S-module, for some ring S. If a function $f \in R^M$ satisying $f(0_M) = 1_R$ is a regular R-valued S-submodule of M then surely $f(s_1 m_1 + s_2 m_2) \succeq f(m_1)f(m_2)$ for all $m_1, m_2 \in M$ and $s_1, s_2 \in S$. Conversely, if this condition holds then, taking $s_1 = s_2 = 1_S$ we see that $f(m_1 + m_2) \succeq f(m_1)f(m_2)$ for all $m_1, m_2 \in M$. Also, we see that $f(sm) = f(sm + 1_S 0_M) \succeq f(m)f(0_M) = f(m)$ for all $s \in S$ and $m \in M$, so f is a regular R-valued S-submodule of M.

(7.1) PROPOSITION. [426] *Let R be a semiring. Let $(M,+)$ be a left S-module for some ring S and let $f \in R^M$ be an R-valued S-submodule of M satisfying the condition that $f(m) \preceq 1$ for all $m \in M$. Then $M_f = \{m \in M \mid f(m) = 1\}$, if nonempty, is an S-submodule of M.*

PROOF. If $m_1, m_2 \in M_f$ then $f(m_1 + m_2) \succeq f(m_1)f(m_2) = 1$ and so, by hypothesis, $f(m_1 + m_2) = 1$. Similarly, if $m \in M$ and $s \in S$ then $f(sm) \succeq f(m) = 1$ and so $f(sm) = 1$. □

Let S be a ring and let M be a left S-module. If R is a multiplicatively-idempotent semiring then we see easily see that $f \in R^M$ is an R-valued S-submodule of M if and only if $L_r(f)$ is an R-valued S-submodule of M for each $r \preceq f(0_M)$.

Note that, as with R-valued subgroups of a group, if f and g are R-valued S-submodules of a left S-module M then $f+g$ need not be an R-valued S-submodule of M. Also note the following example, due to [252]: let $M = (\mathbb{Z}, +)$, considered as a left $\mathbb{Z}$-module, and let $R = (\mathbb{I}, \vee, \wedge)$. Define $f, g \in R^M$ as follows:

$$f: k \mapsto \begin{cases} 1 & \text{if } k \in 2\mathbb{Z} \\ 0 & \text{otherwise} \end{cases}$$

and

$$g: k \mapsto \begin{cases} 1 & \text{if } k \in 3\mathbb{Z} \\ 0 & \text{otherwise.} \end{cases}$$

Then f and g are R-valued regular $\mathbb{Z}$-submodules of M but $h = f \vee g$ is not a $\mathbb{Z}$-submodule of M, for $0 = h(1) = h(3-2) < h(3) \wedge h(2) = 1$.

However, if R is a complete simple additively-idempotent semiring and if $f, g \in R^M$ and subadditive R-valued S-submodules of a a left S-module M then $f + g$ is an R-valued S-submodule of M.

Recall that if $f \in R^M$ is an R-valued S-submodule and if $s \in S$ then define the function $sf \in R^M$ by setting $sf: m \mapsto \sum_{sm'=m} f(m')$. Thus, if $p_{m,r} \in pt(R^M)$ and if $s \in S$ then $sp_{m,r} = p_{sm,r}$.

If $m_1, m_2 \in M$ and $f \in R^M$ is a subadditive R-valued S-submodule then

$$sf(m_1 + m_2) = \sum \{f(m') \mid sm' = m_1 + m_2\}$$
$$\succeq \sum \{f(m_1' + m_2') \mid sm_1' = m_1 \text{ and } sm_2' = m_2\}$$
$$\succeq \sum \{f(m_1') + f(m_2') \mid sm_1' = m_1 \text{ and } sm_2' = m_2\}$$
$$= \sum \{f(m_1') \mid sm_1' = m_1\} + \sum \{f(m_2') \mid sm_2' = m_2\}$$
$$= sf(m_1) + sf(m_2).$$

Moreover, if S is commutative and $s' \in S$ then for each $m \in M$ we have

$$(sf)(m) = \sum \{f(m') \mid sm' = m\}$$
$$= \sum \{f(m') \mid s'sm' = s'm\}$$
$$= \sum \{f(m') \mid s(s'm') = s'm\}$$
$$\preceq \{f(s'm') \mid s(s'm') = s'm\}$$
$$\preceq \{f(m'') \mid sm'' = s'm\}$$
$$= sf(s'm)$$

and so, if S is commutative then $sf \in R^M$ is a subadditive R-valued S-submodule of M.

As a consequence of Proposition 5.21 we note that if R is a complete simple additively-idempotent semiring and if M is a left S-module, where S is a commutative ring, then the family of all subadditive R-valued S-submodules of M is itself a left S-module. Moreover, for $s \in S$ and for subadditive R-valued left S-submodules f, g of M we have $s(f\langle + \rangle g) = sf\langle + \rangle sg$.

(7.2) PROPOSITION. *Let S be a ring and M be a left S-module. If R is a commutative difference-ordered semiring satisfying the condition that (R, M) is a convolution context and if $f, g \in R^M$ are R-valued S-subsemimodules of M, then so is $f\langle + \rangle g$.*

PROOF. By the Corollary to Proposition 5.8, we see that $f\langle+\rangle g$ is an R-valued subsemigroup of M. If $m \in M$ then

$$\begin{aligned}(f\langle+\rangle g)(m) &= \sum_{m'+m''=m} f(m')g(m'') \\ &= \sum_{m+m'=m} f(-m')g(-m'') \\ &= \sum_{m'+m''=-m} f(m')g(m'') \\ &= (f\langle+\rangle g)(-m).\end{aligned}$$

Finally, if $a \in S$ and $m \in M$ then whenever $m'+m'' = m$ we have $f(m')g(m'') \preceq f(am')g(am'')$ so

$$\sum_{m'+m''=m} f(m')g(m'') \preceq \sum_{x'+x''=am} f(x')g(x'').$$

Thus $f\langle+\rangle g$ is an R-valued S-semimodule of M. □

(7.3) PROPOSITION. *Let R be a commutative semiring and let M be a left S-module for some ring S. If $f, g \in R^M$ are [regular] R-valued S-submodules of M then so is fg.*

PROOF. This is a direct consequence of Proposition 6.11. □

(7.4) PROPOSITION. *Let R be an CLO-semiring and let M be a left S-module for some ring S. If U is a nonempty set of [regular] R-valued S-semimodules of M then so is $\wedge U$.*

PROOF. By Proposition 5.12, we know that $\wedge U$ is a [regular] submonoid of $(M, +)$. If $m \in M$ and $a \in S$ then

$$(\wedge U)(am) = \wedge\{f(am) \mid f \in U\} \succeq \wedge\{f(m) \mid m \in U\} = (\wedge U)(m)$$

which shows that $\wedge U$ is an R-valued S-subsemimodule of M, which is surely regular if each element of U is regular. □

Let R be a CLO-semiring and let M be a a left S-module, for some ring S. Then the function $f_1 \in R^M$ defined by $f_1 \colon m \mapsto 1$ for all $m \in M$ is a regular R-valued S-submodule of M which clearly satisfies the condition that $f_1 \succeq g$ for any $g \in R^M$. If U is a nonempty set of R-valued [regular] S-submodules of M then, by Proposition 7.4, $\wedge U$ is also a [regular] R-valued S-submodule of M. Therefore, given any set G of functions from M to R, we can consider

$$\wedge\{f \in R^M \mid f \text{ is an } R\text{-valued [regular] } S\text{-submodule of } M \text{ and } f \succeq g \,\forall g \in G\}.$$

This is a [regular] R-valued S-submodule of M, called the [regular] R-valued S-submodule of M *generated* by G.

For example, let R be a semiring and let M be a a left S-module, for some ring S. Let $p_{m,r} \in pt(R^M)$, where $r \in R$ satisfies $r \succeq r^2$, and let $f_{m,r} \in R^M$ be the function defined by

$$f_{m,r}: m' \mapsto \begin{cases} r & \text{if } m' \in Sm \\ 0 & \text{otherwise} \end{cases}.$$

Thus, if R is an additively-idempotent complete semiring with necessary summation then $f_{m,r} = \sum_{s \in S} p_{sm,r}$ for all $m \in M$ and all $r \in R$.

Then, as we have seen, $f_{m,r}$ is an R-valued S-submodule of M and $f_{m,r} \succeq p_{m,r}$. If $g \in R^M$ is an R-valued S-submodule of M satisfying $g \succeq p_{m,r}$ then

$$g(sm) \succeq g(m) \succeq p_{m,r}(m) = r$$

for each $s \in S$ and so $g \succeq f_{m,r}$. Thus $f_{m,r}$ is the R-valued S-submodule of M generated by $\{p_{m,r}\}$.

Now assume that R is additively idempotent and let $U = \{p_{m_1,r_1}, \ldots, p_{m_t,r_t}\}$ be a finite subset of $pt(R^M)$, where $r_i \succeq r_i^2$ for each $1 \leq i \leq t$. Set $f_U = \sum_{i=1}^{t} f_{m_i,r_i}$. Then $f_U \succeq p_{m_i,r_i}$ for each $1 \leq i \leq t$. Moreover, if $g \in R^M$ is an R-valued S-submodule of M satisfying $g \succeq p_{m_i,r_i}$ for each $1 \leq i \leq t$ then, by additive idempotence, $g \succeq \sum_{i=1}^{t} f_{m_i,r_i} = f_U$. Therefore f_U is the R-valued S-submodule of M generated by U. An R-valued S-submodule of M of the form f_U is *generated by finiely-many points*. Note that even if M is a finitely-generated left S-module it does not necessarily follow that every $f \in R^M$ is generated by finitely-many points. Yu [414] gives the following example: let $M = \mathbb{R}^2$ and let R be the semiring $(\mathbb{I}, \vee, *)$ where $*$ is the triangular norm defined by $a * b = 0 \vee (a + b - 1)$. Then the constant function $f \in R^M$ given by $f: m \mapsto \frac{1}{2}$ for all $m \in M$ is an R-valued $\mathbb{R}$-submodule of M which is not generated by finitely-many points.

In [140] we constructed theories of decomposition and dimension on module categories with values in complete distributive lattices. These constructions can be translated in a straightforward manner to take values in complete semirings. Thus, for example, let R be a complete semiring and let S be a ring. An R-valued *quasidimension function* on the category of all left S-modules is a rule Δ which assigns to each left S-module M an element $\Delta(M)$ of R satisfying the following conditions:

(1) If $(0) \to M' \to M \to M'' \to (0)$ is an exact sequence of left S-modules then $\Delta(M) = \Delta(M') + \Delta(M'')$.
(2) If M is the directed union of a directed family $\{M_1 \mid i \in \Omega\}$ of submodules then $\Delta(M) = \sum_{i \in \Omega} \Delta(M_i)$.

If, in addition, we require that $\Delta(M) \neq 0$ if $M \neq (0)$ then Δ is said to be a *predimension function*. Several important examples of such theories with values in various complete distributive lattices are given in [140]. Thus, for example, if S is a ring and if $S - tors$ is the complete distributive lattice of all (hereditary) torsion theories on the category $S - mod$ of left S-modules, then the rule $supp(\_)$ which assigns to each left S-module its torsion-theoretic support is a quasidimension function which is a predimension function if the ring S is left semidefinite. Similarly, the rule $Spd(\_)$ which assigns to each left S-module its strong projective dimension, in the sense of [311], is a quasidimension function with values in $\mathbb{N} \cup \{\infty\}$.

(7.5) PROPOSITION. *Let S be a ring and let R be a complete semiring. Let Δ be an R-valued quasidimension function on the category of all left S-modules and let $\{M_i \mid i \in \Omega\}$ be a set of left D-modules. Then $\Delta(\coprod_{i \in \Omega} M_i) = \sum_{i \in \Omega} \Delta(M_i)$.*

PROOF. Let $M = \coprod_{i \in \Omega} M_i$. Then M is a directed union of its submodules of the form $\coprod_{i \in \Lambda} M_i$, where Λ ranges over all finite subsets of Ω. Therefore

$$\Delta(M) = \sum \left\{ \Delta(\coprod_{i \in \Lambda} M_i) \,\bigg|\, \Lambda \text{ is a finite subset of } \Delta \right\}.$$

But for each such Λ, $\Delta(\coprod_{i \in \Lambda} M_i) = \sum_{i \in \Lambda} \Delta(M_i)$ by repeated use of condition (1) of the definition of a quasi-dimension function. $\square$

Let R be a commutative simple semiring and let $f \in R^A$, where A is a nonempty set. If S is a ring, then each $a \in A$ defines a function $v_a \in S^{(A)}$ by setting

$$v_a : a' \mapsto \begin{cases} 1 & \text{if } a' = a \\ 0 & \text{otherwise} \end{cases}.$$

Moreover, each element of $S^{(A)}$ has a unique representation in the form $\sum_{a \in A} s_a v_a$, where only finitely-many of the coefficients s_a are nonzero. Now define the function $f' : S^{(A)} \to R$ by setting

$$f' : \sum s_a v_a \mapsto \prod \left\{ f(a) \,\bigg|\, s_a \neq 0 \right\}$$

(where the product over the empty set is taken to be 1). Then f' is an R-valued S-submodule of $S^{(A)}$, called the *free R-valued S-submodule of $S^{(A)}$ generated* by f.

(7.6) PROPOSITION. [291] *Let R be a commutative simple semiring, let S be a ring, and let M be a left S-module. If $g \in R^M$ is an R-valued S-submodule of*

M then there exists a set A and an epimorphism of S-modules $h: S^{(A)} \to M$ such that $g = h[f]$ for some free R-valued S-submodule f of $S^{(A)}$.

PROOF. Pick $A = M$ and let $h: S^{(M)} \to M$ be defined by $h: \sum s_m v_m \mapsto \sum s_m m$. This is surely an epimorphism of left S-modules. Let g' be the free R-valued S-submodule of $S^{(A)}$ generated by g. Then $g = h[g']$. $\square$

R-valued subspaces of a vector space. If S is taken to be a field F, then we obtain the notion of an *R-valued subspace* of a vector space, first studied by Katsaras and Liu [197]. Fuzzy subspaces of vector spaces with values in $(\mathbb{I}, \vee, *)$, where $*$ is a triangular norm, were first studied by Das [73]. Refer also to [255, 261]. These constructions extend the notion of a valued vector space, with valuations in a bounded totally-ordered set. Such constructions are very important, inter alia, in the modern development of infinite abelian group theory, especially in the work of Ulm and its generalizations. From what we noted above, we see that if V is a vector space over a field F and if f is an R-valued subspace of V then $f(av) = f(v)$ for all $0 \neq a \in F$. Thus, if R is a bounded linearly-ordered set so that $(R, \vee, \wedge)$ is a semiring, then we are reduced to the classical case of valuations on a vector space.

Let us look at some more examples.

(IV) EXAMPLE. Let V be a vector space over a field F and let R be a semiring satisfying $1 \succeq r$ for all $r \in R$. Let W be a proper subspace of V and let r be an element of R satisfying $r^2 \preceq r \neq 1$. Then the function $f \in R^V$ defined by

$$f: v \mapsto \begin{cases} 1 & \text{if } v \in W \\ r & \text{otherwise} \end{cases}.$$

is an R-valued regular subspace of V. In particular, the characteristic function on a proper subspace of V (with values in R) is an R-valued regular subspace of V. If R is commutative, then by Proposition 7.3 this means that if f is an R-valued subspace of V and if W is a proper subspace of V then the function $f_W \in R^V$ defined by

$$f_W: v \mapsto \begin{cases} f(v) & \text{if } v \in W \\ 0 & \text{otherwise} \end{cases}$$

is an R-valued subspace of V.

(V) EXAMPLE. Let F be a field. Let Ω be a nonempty set and for each $i \in \Omega$ let V_i be a vector space over F. Set $V = \bigoplus_{i \in \Omega} V_i$. Let R be a commutative semiring and for each $i \in \Omega$ let f_i be an R-valued regular subspace of V. Then we have an R-valued regular subspace f of V defined by: $f: \sum_{i \in \Omega} v_i \mapsto \prod_{i \in \Omega} f_i(v_i)$. This function is well-defined since only finitely-many of the summands v_i are nonzero and since $f_i(0_{V_i}) = 1_R$ by regularity.

(VI) EXAMPLE. The following construction is based on [1]. Let R be a simple additively-idempotent semiring on which we have a complementation δ defined which satisfies the condition that $\delta(r_1 + r_2) \succeq \delta(r_1)\delta(r_2)$ for all $r_1, r_2 \in R$. Let V be a vector space over a field F and let $D(V) = Hom_F(V, F)$. be the dual space of V.

If $f \in R^V$ is an R-valued subspace of V, we define a function $D(f) \in R^{D(V)}$ by setting

$$D(f): \alpha \mapsto \delta\left(\sum_{v \in supp(\alpha)} f(v)\right)$$

(where, as usual, the sum over an empty set of elements of R is taken to be 0_R). We claim that $D(f)$ is an R-valued subspace of $D(V)$. Indeed, if $\alpha, \beta \in D(V)$ then, by Proposition 1.11 and the additive idempotence of R,

$$D(f)(\alpha + \beta) = \delta\left(\sum_{v \in supp(\alpha+\beta)} f(v)\right)$$

$$= \delta\left(\sum_{v \in supp(\alpha) \cup supp(\beta)} f(v)\right)$$

$$= \delta\left(\sum_{v \in supp(\alpha)} f(v) + \sum_{y \in supp(\beta)} f(y)\right)$$

$$\succeq \delta\left(\sum_{v \in supp(\alpha)} f(v)\right) \delta\left(\sum_{y \in supp(\beta)} f(y)\right)$$

$$= D(\alpha)D(\beta).$$

Moreover,

$$D(f)(-\alpha) = \delta\left(\sum_{v \in supp(-\alpha)} f(v)\right) = \delta\left(\sum_{v \in supp(\alpha)} f(v)\right) = D(f)(\alpha)$$

and if $0_R \neq a \in R$ then $supp(\alpha) = supp(a\alpha)$ since R is zerosumfree and so

$$D(f)(a\alpha) = \delta\left(\sum_{v \in supp(a\alpha)} f(v)\right)$$

$$= \delta\left(\sum_{v \in supp(\alpha)} f(v)\right)$$

$$= D(f)(\alpha).$$

(7.7) PROPOSITION. *Let V be a vector space over a field F and let R be a zerosumfree semiring. A subset W of V satisfies the condition that its characteristic function $\chi_W: V \to R$ is an R-valued subspace of V if and only if W is a subspace of V.*

PROOF. This is an immediate consequence of Proposition 5.1. □

Let f be a regular R-valued subspace of a vector space V over a field F. A finite set $D = \{v_1, \ldots, v_n\}$ of elements of V is *linearly independent in f* if and only if

(1) D is linearly independent over F (in the classical sense);
(2) If $a_1, \ldots, a_n$ are elements of F then $f(\sum_{i=1}^n a_i v_i) = \prod_{i=1}^n f(a_i v_i)$.

An arbitrary set D of elements of V is *linearly independent in f* if and only if every finite subset of D is linearly independent in f. A basis for V which is linearly independent in f is a *basis in f*.

In particular, we note that if f is a regular R-valued subspace of a vector space V over a field F and if $0_V \neq v \in V$ then $\{v\}$ is linearly independent in f. Also, let V be a vector space over a field F. If $f \in R^V$ is an R-valued subspace of V and if D is a nonempty subset of V which is linearly independent in f and satisfying the condition that there exists an element $r_0 = r_0^2 \in R$ satisfying $f(v) = r_0$ for all $v \in D$ then for any choice of $v_1, \ldots, v_n \in D$ and $a_1, \ldots, a_n \in F$ we have

$$f\left(\sum_{i=1}^n a_i v_i\right) = \prod_{i=1}^n f(v_i) = r_0^n = r_0$$

and so $f(v) = r_0$ for each vector v in the subspace FD of V spanned by D.

The converse is also true. Let D be a nonempty linearly-independent subset of a vector space V over a field F and assume that there exists a regular R-valued subspace f of V satisfying the condition that $f(v) = r_0$ for some $r_0 = r_0^2 \in R$ and all $v \in FD$. then for any $\{v_1, \ldots, v_n\} \subseteq D$ and any $a_1, \ldots, a_n \in F$ we have

$$f\left(\sum_{i=1}^n a_i v_i\right) = r_0 = r_0^n = \prod_{i=1}^n f(a_i v_i)$$

and so D is linearly independent in f.

Let $F = \mathbb{R}$ and $V = F^2$. Define the $\mathbb{I}$-valued subspace f of V by setting

$$f: v \mapsto \begin{cases} 1 & \text{if } v = (0,0) \\ \frac{1}{2} & \text{if } v = (0, a) \text{ for some } 0 \neq a \in F \\ \frac{1}{4} & \text{otherwise.} \end{cases}$$

Then the subset $D = \{(1,0), (-1,1)\}$ of V is linearly independent but not linearly independent in f. See [255].

Similarly, let R be a semiring and define the R-valued subspace f of V by setting
$$f: (a,b) \mapsto \begin{cases} 1 & \text{if } a = b \\ 0 & \text{otherwise} \end{cases}.$$

Then the standard basis $D = \{(1,0), (0,1)\}$ of V is not linearly independent in f since $f(1,1) = 1 \neq 0 = f(1,0)f(0,1)$ [291].

Consider the following example [255]: let V be a vector space over a field F having a basis $D = \{v_i \mid i \in \Omega\}$ and let $0 \neq b_0 \in \mathbb{I}$. For each $i \in \Omega$ pick $0 \neq b_i \in \mathbb{I}$ satisfying $b_0 \succeq b_i$ for all $i \in \Omega$. Let $f \in \mathbb{I}^V$ be the function defined in the following way:

(1) $f(0_V) = 1$;
(2) $f(v_i) = b_i$ for all $i \in \Omega$;
(3) If $0_V \neq v \in V$ can be writen in a unique manner as $\sum_{j=1}^k c_j v_{i_j}$ with $0 \neq c_j \in F$ then set $f(v) = \wedge_{j=1}^k b_{i_j}$.

Then f is a $\mathbb{I}$-valued subspace of V and D is a basis for V in f.

(7.8) PROPOSITION. *Let R be a commutative semiring and let V be a vector space over a field F. Let f be a regular R-valued subspace of V and let D be a nonempty set of nonzero vectors in V satisfying the condition that if $v_0 \in D$ then $f(v_0) \neq \prod_{v \in B} f(v)$ for any finite subset B of $D \setminus \{v_0\}$. Then D is linearly independent.*

PROOF. By the definition of linear independence in f it suffices to consider the case that D is finite, say $D = \{v_1, \ldots, v_n\}$. We now proceed by induction on n. If $n = 1$ the result is immediate. Assume therefore that $n > 1$ and that the result has already been established for $n - 1$. Then, in particular, $\{v_1, \ldots, v_{n-1}\}$ is linearly independent. If D is not linearly independent then there exists a finite subset T of $\{1, \ldots, n-1\}$ and nonzero scalars $\{a_j \mid j \in T\}$ such that $v_n = \sum_{j \in T} a_j v_j$. Therefore $f(v_n) = \prod_{j \in T} f(a_j v_j) = \prod_{j \in T} f(v_j)$, which contradicts the hypothesis. Therefore D is linearly independent. $\square$

In particular, let R be a commutative semiring and let V be a vector space over a field F. Let f be a regular R-valued subspace of V and let $D = \cup_{i \in \Omega} D_i \subseteq V$, where each D_i is a nonempty linearly-independent subset of V such that the following conditions are satisfied:

(1) If $i \in \Omega$ then there exists an element $r_i = r_i^2$ of R satisfying $f(v) = r_i$ for all $v \in FD$;
(2) If $i \in R$ then $r_i \neq \prod_{j \in \Lambda} r_j$ for any finite subset Λ of $\Omega \setminus \{i\}$.

Then D is linearly independent. This result was used to considerable effect for the case $R = (\mathbb{I}, \vee, \wedge)$ in [254].

(7.9) PROPOSITION. *Let R be a commutative difference-ordered semiring and let V be a vector space over a field F. Let f be a regular R-valued subspace of V and let $v, w \in V$ be vectors satisfying the following conditions:*

(1) $f(v)f(w) = f(v)$;
(2) $f(v)$ *is prime; and*
(3) $f(v) \not\succeq f(w)$.

Then, for all $a, b \in F \setminus \{0\}$, the subset $\{av + bw, v\}$ of V is linearly independent and $f(av + bw) = f(v)$.

PROOF. By Proposition 6.6 we note that

$$f(av + bw) = f(av)f(bw) = f(v)f(w) = f(v)$$

and so all we are left to show is linear independence. Indeed, if there exist $c, d \in F \setminus \{0\}$ such that $c(av + bw) + dv = 0_V$ then the subset $\{v, w\}$ of V is linearly dependent and so, by Proposition 7.8, $f(v) = f(w)$. This contradicts (3) and so we must have linear independence. $\square$

Let R be a complete semiring and let V be a vector space over a field F. Let U be the subset of R^V consisting of all regular R-valued subspaces of V. From Chapter 2, we know that there exists a linear closure operator $f \mapsto f^c$ on R^V defined by

$$f^c = \bigwedge \{g \in U \mid g \succeq f\}.$$

In particular, if $\varnothing \neq A \subseteq V \setminus \{0_V\}$ and if $f_A \in R^V$ is the function defined by

$$f_A : v \mapsto \begin{cases} \infty & \text{if } 0_V \neq v \in A \\ 0 & \text{otherwise} \end{cases}$$

then we see that whenever $g \in U$ and $g \succeq f_A$ we must have $g(0_V) = 1_R$ and $g(w) = \infty$ for all $0_V \neq w \in FA$. Therefore f^c is given by

$$f_A^c : v \mapsto \begin{cases} \infty & \text{if } 0_V \neq v \in FA \\ 1_R & \text{if } v = 0_V \\ 0 & \text{otherwise} \end{cases}.$$

In particular, f_A is c-closed if and only if $supp(f_A) \cup \{0_V\}$ is a subspace of V. If $y \in A$ then

$$f \neg p_{y,1} : v \mapsto \begin{cases} \infty & \text{if } v \in A \setminus \{y\} \\ 0_V & \text{otherwise} \end{cases}$$

so $supp((f \neg p_{y,1})^c) = F(A \setminus \{y\})$. Thus f_A is c-free if and only if $y \notin F(A \setminus \{y\})$ for all $y \in A$. In other words, f_A is c-free if and only if A is free in the usual sense of vector-space theory. Similarly, f_A is a c-basis for f_A^c if and only if A is a basis for the subspace FA of V.

Given a vector space V over a field F and a regular R-valued subspace f of V, it is not necessarily true that V has a basis in f. For example [2]: let F be a field and let $V = F^\infty$ be the vector space of all countable sequences $[a_1, a_2, \ldots]$ of elements of F. Let $f \in \mathbb{I}^V$ be the $\mathbb{I}$-valued subspace of V defined as follows:

$$f: [a_1, a_2, \ldots] \mapsto \begin{cases} 1 & \text{if } a_i = 0 \text{ for all } i \geq 0 \\ 1 - \frac{1}{h+1} & \text{if } a_1 = \cdots = a_{h-1} = 0 \text{ but } a_h \neq 0 \end{cases}.$$

Then V has no basis in f.

(7.10) PROPOSITION. *Let R be a semiring and let V be a vector space over a field F. Let f be a regular R-valued subspace of V and assume that there exists a finite basis $\{v_1, \ldots, v_n\}$ in f. Then $im(f)$ is finite.*

PROOF. If $v \in V$ then there exists scalars $a_1, \ldots, a_n$ in F satisfying $v = \sum_{i=1}^n a_i v_i$ and so $f(v) = f(\sum_{i=1}^n a_i v_i) = \prod_{i=1}^n f(a_i v_i) = \prod_{i=1}^n t_i$, where $t_i = f(v_i)$ or $t_i = f(0_V)$. The set of all elements of R of this form is surely finite. $\square$

We conclude therefore that if R is a semiring and if V is a vector space over a field F having a regular R-valued subspace f the image of which is infinite, then there is no basis in f.

Lybczonok [255] proves that a vector space V over a field F has a basis in a $\mathbb{I}$-valued subspace f whenever $im(f)$ is closed under taking arbitrary suprema. In particular, this means that if V is finitely-generated then V has a basis in every $\mathbb{I}$-valued subspace. This result was extended in [2] to show that if V is countably-generated then V has a basis in every $\mathbb{I}$-valued subspace.

(7.11) PROPOSITION. [255] *Let R be a complete multiplicatively-idempotent semiring. Let f be a regular R-valued subspace of a vector space V over a field F satisfying the property*

(*) *If $\varnothing \neq D \subseteq im(f)$ then $\sum D \in D$.*

If W is a proper subspace of V then there exists an element y_0 of $V \setminus W$ satisfying $f(y_0 + w) f(w) = f(y_0) f(w)$ for all $w \in W$ satisfying $f(y_0) \succeq f(w)$ or $f(w) \succeq f(y_0)$.

PROOF. By hypothesis, there exists an element y_0 of $V \setminus W$ satisfying

$$f(y_0) = \sum \{f(v) \mid v \in V \setminus W\}.$$

Since $y_0 + w \in Y \setminus W$ for each $w \in W$ we then have $f(y_0) \succeq f(y_0 + w)$ for each $w \in W$.

Now let $w \in W$. If $f(y_0) \succeq f(w)$ then

$$f(y_0)f(w) \succeq f(w)f(w)$$
$$= f(w)$$
$$= f(-y_0 + [y_0 + w])$$
$$\succeq f(-y_0)f(y_0 + w)$$
$$= f(y_0)f(y_0 + w)$$
$$\succeq f(y_0)f(y_0)f(w)$$
$$= f(y_0)f(w)$$

and so $f(y_0)f(w) = f(w)$. Thus $f(y_0) \succeq f(y_0 + w) \succeq f(y_0)f(w)$ so $f(y_0)f(w) \succeq f(y_0 + w)f(w) \succeq f(w)^2 = f(w) = f(y_0)f(w)$, proving that

$$f(y_0 + w)f(w) = f(y_0)f(w).$$

If $f(w) \succeq f(y_0)$ then $f(y_0) \succeq f(y_0 + w) \succeq f(y_0)f(w)$ so

$$f(y_0)f(w) \succeq f(y_0 + w)f(w) \succeq f(y_0)f(w)^2 = f(y_0)f(w).$$

□

Let R be a semiring and let V be a vector space over a field F. A regular R-valued subspace $f \in R^V$ is *simply generated* if and only if

(1) $f = p_{0_V,r}$ for some $r \in R$; or
(2) $f = f_U$, where $U = \{p_{0_V,r}, p_{v_1,r_1}, \ldots, p_{v_t,r_t}\}$ and $\{v_0, \ldots, v_t\}$ is a finite subset of V linearly independent in f.

The following result is based on [414].

(7.12) PROPOSITION. *Let R be an additively-idempotent semiring and let V be a vector space over field F. Let $f \in R^V$ be a regular R-valued subspace of V generated by finitely-many points. Then $f = \sum_{j=1}^{m} g_j$, where $\{g_1, \ldots, g_m\}$ is a set of simply-generated regular R-valued subspaces of V.*

PROOF. By hypothesis, $f = f_U$, where $U = \{p_{x_1,r_1}, \ldots, p_{x_t,r_t}\}$. Set $r_0 = \sum_{i=1}^{t} r_i$. If $x_i = 0_V$ for all $1 \leq i \leq t$ then $f = p_{0_V,r_0}$ and we are done. Otherwise, the set $D = \{x_1, \ldots, x_t\}$ has finitely-many maximal linearly-independent subsets, say $D_1, \ldots, D_m$, where each D_j is of the form $\{x_{h(j,1)}, \ldots, x_{h(j,k_j)}\}$ with

$$1 \leq h(j,1) < \cdots < h(j,k_j) \leq t.$$

For each $1 \leq j \leq m$, let

$$U_j = \{p_{0_V,r_0}, p_{x_{h(j,1)},r_{h(j,1)}}, \ldots, p_{x_{h(j,k_j)},r_{h(j,k_j)}}\}$$

and let $g_j = f_{U_j}$. Then each g_j is simply-generated and it suffices to show that $f = \sum_{j=1}^m g_j$. Indeed, set $g = \sum_{j=1}^m g_j$. Clearly $g_j \preceq f$ for all $1 \leq j \leq m$ and so $g \preceq f$ by additive idempotence of R^V. On the other hand, let $v \in V$. If $f(v) = 0$ then surely $f(v) \preceq g(v)$. If $v = 0_V$ then $f(v) = r_0 = g(v)$. Therefore we can assume that $v \neq 0_V$ and that $f(v) \neq 0$.

We have previously noted that $f_U = \sum_{i=1}^t f_{x_i, r_i}$, where

$$f_{x_i, r_i}: y \mapsto \begin{cases} r_i & \text{if } y \in Fx_i \\ 0 & \text{otherwise} \end{cases}.$$

Therefore there exist indices $1 \leq i_1 < \ldots i_p \leq t$ and nonzero scalars $a_{i_1}, \ldots, a_{i_p}$ in F such that $v = \sum_{h=1}^p a_{i_h} x_{i_h}$ and $f(v) = \prod_{h=1}^p r_{i_h}$. Indeed, without loss of generality we can assume that the set $\{x_{i_1}, \ldots, x_{i_p}\}$ is linearly independent. But then this set is contained in D_j for some $1 \leq j \leq m$ and so $f(v) \preceq g_j(v) \preceq g(v)$. Thus $f \preceq g$ and so we have equality. □

This result suggests that it is important to study the simply-generated regular R-valued subspaces of a vector space. For the case of $R = \mathbb{I}$, such a study is carried out in [414].

8. Semiring-valued Ideals in Semirings and Rings

Let $(R, +, \cdot)$ and $(S, \oplus, \odot)$ be semirings. An *R-valued left* [resp. *right*] *ideal* of S is a R-valued subsemigroup f of $(S, \oplus)$ which is not a constant function and which satisfies the following additional condition:

(*) $f(s' \odot s) \succeq f(s)$ [resp. $f(s \odot s') \succeq f(s)$] for all $s, s' \in S$.

Note that in this case,

$$f(0_S) = f(0_S \odot s) \succeq f(s) = f(s \odot 1_S) \succeq f(1_S)$$

[resp. $f(0_S) = f(0_S \odot s) \succeq f(s) = f(1_S \odot s) \succeq f(1_S)$] for all $s \in S$ and so, as a subsemigroup of $(S, \oplus)$, every R-valued left ideal of S is a facile R-valued submonoid of $(S, \oplus)$. The same is true for every R-valued right ideal of S. If $f(0_S) = 1_R$ then f is *regular*. If $(S, \oplus, \odot)$ is a ring then we will also insist that f be a R-valued subgroup of $(S, \oplus)$. A function which is both a [regular] R-valued left ideal and a [regular] R-valued right ideal of S is a [regular] R-valued *ideal* of S.

As usual, we begin with some examples.

(I) EXAMPLE. [95] Consider the semirings $R = (\mathbb{I}, \vee, \wedge)$ and $S = (\mathbb{N}, +, \cdot)$. Then the function $f \in R^S$ defined by

$$f : k \mapsto \begin{cases} 1 & \text{if } k = 0 \\ \frac{1}{2} & \text{if } 0 \neq k \in 2\mathbb{N} \\ \frac{1}{3} & \text{otherwise} \end{cases}$$

is a regular R-valued ideal of S.

(II) EXAMPLE. [230] Let S be a ring. Any ideal of S can be considered as an $\mathbb{I}$-valued ideal the image of which is in $\{0,1\}$. One can show that the ring S is artinian and noetherian if and only if the image of every $\mathbb{I}$-valued ideal of S is finite. If S is a complete lattice then the $\mathbb{I}$-valued ideals of S have been studied in [243].

(III) EXAMPLE. [194] Let S be a semiring. If $f \in \mathbb{I}^S$ is an $\mathbb{I}$-valued left [right] ideal of S then the function $f^+ \in \mathbb{I}^S$ defined by $f^+ : s \mapsto f(s) + 1_R - f(0)$ is a regular $\mathbb{I}$-valued left [right] ideal of S. Moreover, f is regular if and only if $f = f^+$.

Let $(R,+,\cdot)$ and $(S,\oplus,\odot)$ be semirings. If $f,g \in R^S$ are R-valued left [right] ideals of S, one of which is central, then $f \times g : (s,s') \mapsto f(s)g(s')$ is an R-valued left [right] ideal of $S \times S$ which is regular whenever both f and g are regular. It is possible, however, for $f \times g$ to be an R-valued ideal of S even though both f and g are not. For example, let S be any ring having at least two elements and let $0 \leq t_0 < t_1 < 1$ in $\mathbb{I}$. Let $f,g \in \mathbb{I}^S$ be the the functions defined by $f: s \mapsto t_0$ and

$$g: s \mapsto \begin{cases} t_1 & \text{if } s = 0_S \\ 1 & \text{otherwise} \end{cases}.$$

Then g is not an $\mathbb{I}$-valued ideal of S while f is. Moreover, $f \times g = f$ so $f \times g$ is also an $\mathbb{I}$-valued ideal of S.

(8.1) PROPOSITION. Let $(R,+,\cdot)$ and $(S,\oplus,\odot)$ be semirings and let $f \in R^S$ be central. Then f is a regular R-valued [left, right] ideal of S if and only if $f \times f$ is a regular R-valued [left, right] ideal of $S \times S$.

PROOF. As we have just seen, if f is a regular R-valued [left, right] ideal of S then $f \times f$ is a regular R-valued [left, right] ideal of $S \times S$. Conversely, assume that this is the case. By Proposition 5.3, f is a regular R-valued submonoid of $(S,\oplus)$. If $s \in S$ then $f(-s) = (f \times f)(-s,0) = (f \times f)(s,0) = f(s)$ and so f is also a regular R-valued subgroup of S.

Now assume that f f is an R-valued left ideal of $S \times S$. Then

$$f(s' \odot s) = (f \times f)((s',0) \odot (s,0)) \succeq (f \times f)(s,0) = f(s).$$

The case of right ideals is handled similarly. $\square$

Let R and S be a semirings, with R additively idempotent. Then condition (*) says that if $f \in R^S$ is a [regular] R-valued ideal of R we have $f(ss') \succeq f(s) + f(s')$ for all $s,s' \in S$. Indeed, if S is in fact a ring then this, together with the R-valued subgroup condition, immediately implies that $f \in R^S$ is a [regular] R-valued ideal of S if and only if $f(s - s') \succeq f(s)f(s')$ and $f(ss') \succeq f(s) + f(s')$

for all $s, s' \in S$. We also note that if S is a ring and if $f \in R^S$ is a [regular] R-valued ideal of S then for elements $s, s' \in S$ satisfying $f(s - s') = 1_R$ we have $f(s) = f(s - s' + s') \succeq f(s - s')f(s') = f(s')$ and similarly $f(s') \succeq f(s)$, proving that $f(s) = f(s')$.

(8.2) PROPOSITION. Let $(R, +, \cdot)$ and $(S, \oplus, \odot)$ be semirings, with R additively idempotent, and let $f \in R^S$ be an R-valued ideal of S. Then $H_f = \{s \in S \mid f(s) \succ f(1_S)\}$ is an ideal of S. If R is multiplicatively idempotent as well then $K_f = \{s \in S \mid f(s) = f(0_S)\}$ is also an ideal of S.

PROOF. Since R is additively idempotent, $\preceq$ is a partial order on R. If $s, s' \in H_f$ then $f(s \oplus s') \succeq f(s)f(s') \succeq f(s) \succ f(1_S)$ and so $s \oplus s' \in H_f$. If $s \in H_f$ and $s' \in S$ then $f(s' \odot s) \succeq f(s) \succ f(1_S)$ so $s' \odot s \in H_f$. Similarly, $s \odot s' \in H_f$.

Now assume that R is multiplicatively idempotent as well. If $s, s' \in K_f$ then $f(0_S) \succeq f(s \oplus s') \succeq f(s)f(s') = f(0_S)^2 = f(0_S)$ so $f(s \oplus s') = f(0_S)$, proving that $s \oplus s' \in K_f$. If $s \in H_f$ and $s' \in S$ then $f(0_S) \succeq f(s' \odot s) \succeq f(s) = f(0_S)$, proving that $s' \odot s \in K_f$. Similarly $s \odot s' \in K_f$. $\square$

We note, of course, that in the above situation $K_f \subseteq H_f$.

(8.3) PROPOSITION. Let $(R, +, \cdot)$ and $(S, \oplus, \odot)$ be semirings, with R difference-ordered. Let $f \in R^S$ be a left [right] R-valued ideal of S satisfying the condition that $f(0_S)$ is multiplicatively idempotent. Then $I = \{s \in S \mid f(s) = f(0_S)\}$ is a left [right] ideal of S.

PROOF. If $s, s' \in I$ then $f(0_S) \succeq f(s \oplus s') \succeq f(s)f(s') = f(0_S)^2 = f(0_S)$ and so $s \oplus s' \in I$. If f is an R-valued left ideal of S and if $s' \in S$ and $s \in I$ then $f(0_S) \succeq f(s' \odot s) \succeq f(s) = f(0_S)$ so $s's \in I$. Thus I is a left ideal of S. The proof for right ideals is similar. $\square$

If $(S, \oplus, \odot)$ is a ring and R is a semiring, and if $f, g \in R^S$ are R-valued ideals of S, then $f + g$ need not be an R-valued ideal of S, as the following example, due to [85], shows: let $S = \mathbb{Z}/(6)$ and let $t_0 > t_1 > \cdots > t_4$ in $\mathbb{I}$. Let $f, g \in \mathbb{I}^S$ be the $\mathbb{I}$-valued ideals of S defined as follows:

$$f: s \mapsto \begin{cases} t_0 & \text{if } s = 0 \\ t_2 & \text{if } s = 3 \\ t_3 & \text{otherwise} \end{cases}.$$

$$g: s \mapsto \begin{cases} t_1 & \text{if } s = 0 \\ t_2 & \text{if } s = 2 \text{ or } s = 4 \\ t_4 & \text{otherwise} \end{cases}.$$

Then $f + g$ is not an $\mathbb{I}$-valued ideal of S.

On the other hand, if $(S, \oplus, \odot)$ is a ring and R is a semiring, and if $f, g \in R^S$ are R-valued ideals of S, then $f + g$ may be an R-valued ideal of S even if $f \not\succeq g$ and $g \not\succeq f$, as the following example, also due to [85], makes clear: let $(S, \oplus, \odot)$ be any ring and let $t_0 > t_1 > t_2 > t_3$ in $\mathbb{I}$. Let $f, g \in \mathbb{I}^S$ be the $\mathbb{I}$-valued ideals of S defined by

$$f: s \mapsto \begin{cases} t_0 & \text{if } s = 0 \\ t_3 & \text{otherwise} \end{cases}.$$

$$g: s \mapsto \begin{cases} t_1 & \text{if } s = 0 \\ t_2 & \text{otherwise} \end{cases}.$$

Then $f + g$ is an $\mathbb{I}$-valued ideal of S although $f \not\succeq g$ and $g \not\succeq f$.

(8.4) PROPOSITION. *Let R be a commutative semiring and let $(S, \oplus, \odot)$ be a semiring. If f and g are [regular] R-valued ideals of S then so is the function $fg \in R^S$. If $(R, (S, \odot))$ is a convolution context and if f and g are R-valued ideals of S then so is $f\langle\odot\rangle g$.*

PROOF. First let us consider the function $fg \in R^S$ defined by $fg: s \mapsto f(s)g(s)$. If $s, s' \in S$ then

$$(fg)(s \oplus s') = f(s \oplus s')g(s \oplus s') \succeq f(s)f(s')g(s)g(s')$$
$$= f(s)g(s)f(s')g(s') = (fg)(s)(fg)(s')$$

and, moreover, $(fg)(s' \odot s) = f(s' \odot s)g(s' \odot s) \succeq f(s)g(s) = fg(s)$. Similarly $(fg)(s \odot s') \succeq fg(s)$. Finally, if f and g are both regular then $fg(0_S) = f(0_S)g(0_S) = 1_R 1_R = 1_R$ and so fg is regular. Thus fg is a [regular] R-valued ideal of S.

Now let us consider the function $f\langle\odot\rangle g \in R^S$, where f and g are both assumed to be R-valued ideals of S. By Proposition 6.11, we know that $f\langle\odot\rangle g$ is an R-valued subgroup of $(S, \oplus)$. Moreover, if $f, g \in R^S$ then

$$(f\langle\odot\rangle g)(s' \odot s) = \sum_{x \odot y = s' \odot s} f(x)g(y)$$
$$\succeq \sum_{x \odot y = s} f(x)g(y \odot s)$$
$$\succeq \sum_{x \odot y = s} f(x)g(y)$$
$$= (f\langle\odot\rangle g)(s)$$

and so $f\langle\odot\rangle g$ is an R-valued ideal of S. $\square$

(8.5) Proposition. *Let R be a CLO-semiring with necessary summation and let $(S, \oplus, \odot)$ be a semiring such that $(R, (S, \oplus))$ is a convolution context. If $f, g \in R^S$ are left [right] R-valued ideals of S then so is $f\langle\oplus\rangle g$.*

PROOF. By the corollary to Proposition 5.8 we know that $f\langle\oplus\rangle g$ is an R-valued subsemigroup of $(S, \oplus)$, which is in fact clearly a submonoid. Assume that both f and g are R-valued left ideals of S. Then for $s, s' \in S$ we have

$$\begin{aligned}(f\langle\oplus\rangle g)(s' \odot s) &= \sum_{t' \oplus t'' = s' \odot s} f(t')g(t'') \\ &\succeq \sum_{x \oplus y = s} f(s' \odot x)g(s' \odot y) \\ &\succeq \sum_{x \oplus y = s} f(x)g(y) \\ &= (f\langle\oplus\rangle g)(s)\end{aligned}$$

and so $f\langle\oplus\rangle g$ is an R-valued left ideal of S. The case of right ideals is similar. □

Following the usual definition from ring theory, we say that a semiring $(S, \oplus, \odot)$ is *regular in the sense of Von Neumann* if and only if for each $s \in S$ there exists an element $s' \in S$ satisfying $s = s \odot s' \odot s$.

(8.6) Proposition. *Let R be a commutative additively-idempotent semiring, with necessary summation in case R is complete, and let $(S, \oplus, \odot)$ be a semiring. If f and g are R-valued ideals of S then $f\langle\odot\rangle g \succeq fg$, with equality if S is regular in the sense of Von Neumann.*

PROOF. If $s \in S$ then

$$\begin{aligned}(f\langle\odot\rangle g)(s) &= \sum_{s' \odot s'' = s} f(s')g(s'') \\ &\succeq \sum_{s' \odot s'' = s} f(s' \odot s'')g(s' \odot s'') \\ &= \sum_{s' \odot s'' = s} f(s)g(s) \\ &= f(s)g(s) \\ &= (fg)(s)\end{aligned}$$

and so $f\langle\odot\rangle g \succeq fg$.

Now assume that S is regular in the sense of Von Neumann and let f and g be R-valued ideals of S. If $s \in S$ then

$$(f\langle\odot\rangle g)(s) = \sum_{t' \odot t'' = s} f(t')g(t'') \succeq f(s \odot s')g(s)$$

(where s' is an element of S satisfying $s \odot s' \odot s = s$). But $f(s \odot s') \succeq f(s)$ so we get $(f\langle\odot\rangle g)(s) \succeq f(s)g(s) = (fg)(s)$. Thus we have equality. □

(8.7) PROPOSITION. Let R be a commutative semiring which is both additively and multiplicatively idempotent, with necessary summation in case R is complete, and let $(S, \oplus, \odot)$ be a semiring. Assume that both $(R, (S, \oplus))$ and $(R, (S, \odot))$ are convolution contexts. If $f, g, h \in R^S$ are left [right] R-valued ideals of S then:

(1) $f\langle\odot\rangle(g\langle\oplus\rangle g) \preceq (f\langle\odot\rangle g)\langle\oplus\rangle(f\langle\odot\rangle h)$;
(2) $(f\langle\oplus\rangle g)\langle\odot\rangle h \preceq (f\langle\odot\rangle h)\langle\oplus\rangle(g\langle\odot\rangle h)$.

PROOF. It suffices to prove (1), since the proof of (2) is similar. Ineed, suppose that $s, s', s'' \in S$ satisfy $s = s' \odot s''$. Then:

$$f(s)(g\langle\oplus\rangle h)(s'') = f(s')\left[\sum_{t \oplus t' = s''} g(t)h(t')\right]$$

$$= f(s')^2 \left[\sum_{t \oplus t' = s''} g(t)h(t')\right]$$

$$= \sum_{t \oplus t' = s''} [f(s')g(t)][f(s')h(t')]$$

$$\preceq \sum_{s' \odot t \oplus s' \odot t' = s} [f(s')g(t)][f(s')h(t')]$$

$$\preceq \sum_{s' \odot t \oplus s' \odot t' = s} (f\langle\odot\rangle g)(s' \odot t)(f\langle\odot\rangle h)(s' \odot t')$$

$$\preceq \sum_{u \oplus v = s} (f\langle\odot\rangle g)(u)(f\langle\odot\rangle h)(v)$$

$$= [(f\langle\odot\rangle g)\langle\oplus\rangle(f\langle\odot\rangle h)](s)$$

and so, by additive idempotence and necessary summation,

$$[f\langle\odot\rangle(g\langle\oplus\rangle g)](s) \preceq [(f\langle\odot\rangle g)\langle\oplus\rangle(f\langle\odot\rangle h)](s).$$

Thus we have (1). □

(8.8) PROPOSITION. Let $(R, +, \cdot)$ be a simple lattice-ordered semiring and let $(S, \oplus, \odot)$ be a semiring such that $(R, (S, \oplus))$ and $(R, (S, \odot))$ are convolution contexts. A function $f \in R^S$ satisfying $f(0_S) \succeq f(a)$ for all $a \in S$ [resp. $f(0_S) = 1_R$] is a [regular] R-valued left ideal of S if and only if, for all $r, t \in R$ and all $a, b \in S$ the following two conditions are satisfied:

(1) $p_{a,r}\langle\oplus\rangle p_{b,t} \succeq f$ whenever $p_{a,r}, p_{b,t} \succeq f$;
(2) $p_{a,r}\langle\odot\rangle f \succeq f$.

PROOF. First assume that f is an R-valued left ideal of S. If $p_{a,r}, p_{b,t} \succeq f$ then, for all $z \in S$,

$$[p_{a,r}\langle\oplus\rangle p_{b,t}](z) = \sum_{x\oplus y=z} p_{a,r}(x)p_{b,t}(y) \succeq \sum_{x\oplus y=z} f(x)f(y) = (f\langle\oplus\rangle f)(z)$$

By Proposition 5.8, $(f\langle\oplus\rangle f)(z) \succeq f(z)$ for all $z \in S$ and so $p_{a,r}\langle\oplus\rangle p_{b,t} \succeq f$. Moreover, for all $z \in S$,

$$[p_{a,r}\langle\odot\rangle f](z) = \sum_{x\odot y=z} p_{a,r}(x)f(y) \succeq \sum_{x\odot y=z} f(x)f(y) = (f\langle\odot\rangle f)(z)$$

and so, gain by Proposition 5.8, $p_{a,r}\langle\odot\rangle f \succeq f$.

Now, conversely, assume that conditions (1) and (2) hold. If $a, b \in S$ then

$$f(a \oplus b) \succeq [p_{a,f(a)}\langle\oplus\rangle p_{b,f(b)}](a \oplus b)$$
$$= \sum_{x\oplus y=a\oplus b} p_{a,f(a)}(x)p_{b,f(b)}(y)$$
$$= f(a)f(b)$$

and so f is a [regular] R-valued submonoid of $(S, \oplus)$. Also,

$$f(a \odot b) \succeq [p_{a,1_R}\langle\odot\rangle f](a \odot b) = \sum_{x\odot y=a\odot b} p_{a,1_R}(a)f(b) \succeq f(b)$$

and so f is an R-valued left ideal of S. □

(8.9) PROPOSITION. *Let $(R, +, \cdot)$ be a simple lattice-ordered semiring and let $(S, \oplus, \odot)$ be a semiring such that $(R, (S, \odot))$ is a convolution context. If $f, g \in R^S$ are R-valued ideals of S then $f\langle\odot\rangle g \succeq fg$.*

PROOF. If $a \in S$ then $(f\langle\odot\rangle g)(a) = \sum_{a=x\odot y} f(x)g(y)$. Since f and g are both R-valued ideals of S, we know that $f(x) \succeq f(x \odot y) = f(a)$ and $g(y) \succeq g(x \odot y) = g(a)$ whenever $a = x \odot y$ and so $(f\langle\odot\rangle g)(a) \succeq f(a)g(a) = (fg)(a)$. □

Now let R be a semiring. Let S be a semiring and let M be a left S-semimodule. If $f \in R^S$ is an R-valued ideal of S and if $g \in R^M$ is an R-valued S-subsemimodule of M then, following [7], we define the function $f \bullet g \in R^M$ by

$$f \bullet g : m \mapsto \sum \left\{ f(a_1)g(x_1) \cdot \ldots \cdot f(a_t)g(x_t) \,\middle|\, m = \sum_{i=1}^{t} a_i x_i \text{ where } a_i \in S \text{ and } x_i \in M \right\}.$$

Straightforward calculation shows that $f \bullet g$ is an R-valued S-submodule of M. In particular, if both f and g are R-valued ideals of S then so is $f \bullet g$.

(8.10) PROPOSITION. Let R be a commutative semiring, with necessary summation in case R is complete, in which both addition and multiplication are idempotent. Then the following conditions on a semiring S are equivalent:

(1) $I^2 = I$ for all ideals I of S;
(2) $f \bullet f = f$ for all R-valued ideals f of S;
(3) $f\langle \cdot \rangle g = f \bullet g$ for all R-valued ideals f and g of S.

PROOF. (1) $\Rightarrow$ (2): Let f be an R-valued ideal of S. For any $s \in S$ we have

$$(f \bullet f)(s) = \sum \left\{ f(a_1)f(b_1) \cdot \ldots \cdot f(a_t)f(b_t) \,\middle|\, s = \sum_{i=1}^{t} a_i b_i \right\}$$

$$\preceq \sum \left\{ f(a_1 b_1)f(a_1 b_1) \cdot \ldots \cdot f(a_t b_t)f(a_t b_t) \,\middle|\, s = \sum_{i=1}^{t} a_i b_i \right\}$$

$$= \sum \left\{ [f(a_1 b_1) \cdot \ldots \cdot f(a_t b_t)] \cdot [f(a_1 b_1) \cdot \ldots \cdot f(a_t b_t)] \,\middle|\, s = \sum_{i=1}^{t} a_i b_i \right\}$$

$$\preceq \sum \left\{ f(s)f(s) \,\middle|\, s = \sum_{i=1}^{t} a_i b_i \right\}$$

$$= f(s)$$

Thus $f \bullet f \preceq f$. On the other hand, If $s \in S$ then the ideal (s) of S generated by s is idempotent and so $s \in (s)^2 = SsSSsS$. This means that s can be written in the form $\sum_{i=1}^{t} a_i s a'_i b_i s b'_i$, where the $a_i, a'_i, b_i, b'_i \in S$ for all $1 \leq i \leq t$. Moreover, for each $1 \leq i \leq t$ we have $f(s) = f(s)f(s) \preceq f(a_i s a'_i)f(b_i s b'_i)$ and so

$$f(s) \preceq f(a_1 s a'_1)f(b_1 s b'_1) \cdot \ldots \cdot f(a_t s a'_t)f(b_t s b'_t)$$

$$\preceq \sum \left\{ f(a_1 s a'_1)f(b_1 s b'_1) \cdot \ldots \cdot f(a_t s a'_t)f(b_t s b'_t) \,\middle|\, s = \sum_{i=1}^{t} a_i s a'_i b_i s b'_i \right\}$$

$$\preceq \sum \left\{ f(y_1)f(z_1) \cdot \ldots \cdot f(y_r)f(z_r) \,\middle|\, s = \sum_{i=1}^{r} y_i z_i \right\}$$

$$= (f \bullet f)(s)$$

and so $f = f \bullet f$.

(2) $\Rightarrow$ (1): Let I be an ideal of S and let $f = \chi_I$ be the characteristic function on I, which is an R-valued ideal of S. Then $f \bullet f = f$ and so $I = f^{-1}(1) = (f \bullet f)^{-1}(1) = I^2$, proving (1).

(1) ⇒ (3): Let f and g be R-valued ideals of S. Then for any $s \in S$ we have

$$(f \bullet g)(s) = \sum \left\{ f(y_1)g(z_1) \cdot \ldots \cdot f(y_t)g(z_t) \ \bigg| \ s = \sum_{i=1}^{t} y_i z_i \right\}$$

$$\preceq \sum \left\{ f(y_1 z_1)g(y_1 z_1) \cdot \ldots \cdot f(y_t z_t)g(y_t z_t) \ \bigg| \ s = \sum_{i=1}^{t} y_i z_i \right\}$$

$$= \sum \left\{ [f(y_1 z_1) \cdot \ldots \cdot f(y_t z_t)][g(y_1 z_1) \cdot \ldots \cdot g(y_t z_t)] \ \bigg| \ s = \sum_{i=1}^{t} y_i z_i \right\}$$

$$\preceq \sum \left\{ f(s)g(s) \ \bigg| \ s = \sum_{i=1}^{t} y_i z_i \right\}$$

$$= f(s)g(s) = (fg)(s)$$

and so $f \bullet g \preceq fg$. Again, by the same argument as above, we obtain $fg \preceq f \bullet g$ and so we have equality.

(3) ⇒ (2): This is immediate. □

We note that the product $f \bullet g$ of two R-valued ideals of a semiring S is a special case of the notion of *intrinsic product* first considered in [167, 249]. In general, let M be a set on which we have two operations; $+$ and $*$ defined. Then on R^M we define the operation $\langle\!\langle * \rangle\!\rangle$, called the $*$-*intrinsic product* in the following manner:

(1) If $f, g \in R^M$ and if D is a finite subset of $supp(f) \times supp(g)$ set

$$\sigma(D) = \sum_{(m', m'') \in D} m' * m''$$

and

$$\theta_{fg}(D) = \sum_{(m', m'') \in D} f(m')g(m'');$$

(2) If $m \in M$ set $(f \langle\!\langle * \rangle\!\rangle g)(m) = \sum_{m = \sigma(D)} \theta_{fg}(D)$ where, as usual, the sum of an empty set of summands is taken to be 0_R.

Of course, in order for this sum to make sense, we have to assume that either each element of M can be written as $\sigma(D)$ in only finitely-many ways for each pair $f, g \in R^M$ (which would surely be true if M were finite) or that the ring R is complete. Thus, in our case, if S is a ring we have $\bullet = \langle\!\langle \cdot \rangle\!\rangle$.

(8.11) PROPOSITION. Let R be a QLO-semiring. Let S, S' be rings and let $\gamma: S \to S'$ be a ring homomorphism. If $f \in R^S$ is an R-valued ideal of S then $g = h_\gamma[f]$ is an R-valued ideal of S'. Moreover, if R is simple and f is regular then g is regular.

PROOF. By Proposition 5.14, we know that g is an R-valued subsemigroup of the additive group of S'. Moreover, if $s' \in S'$ then

$$g(s') = \sum_{\gamma(s)=s'} f(s) = \sum_{\gamma(s)=s'} f(-s) = \sum_{\gamma(s)=-s'} f(s) = g(-s')$$

and so g is an R-valued subgroup of the additive group of S'.

If $s'_1, s'_2 \in S'$ then

$$\begin{aligned} g(s'_1 s'_2) &= \sum_{\gamma(s)=s'_1 s'_2} f(s) \\ &\succeq \sum_{\gamma(s_1)=s'_1} \sum_{\gamma(s_2)=s'_2} f(s_1 s_2) \\ &\succeq \sum_{\gamma(s_1)=s'_1} \sum_{\gamma(s_2)=s'_2} f(s_1) \\ &= g(s'_1). \end{aligned}$$

Similarly, $g(s'_1)g(s'_2) \succeq g(s'_2)$ and so g is an R-valued ideal of S'.

If R is simple and f is regular then $1_R = f(0_S) \preceq g(0_{S'}) \preceq 1_R$ and so $g(0_{S'}) = 1_R$, proving that g is regular as well. □

We now turn to the problem of constructing factor rings by regular R-valued ideals. Our approach follows [226].

(8.12) PROPOSITION. Let R be a semiring and let $(S, \oplus, \odot)$ be a ring. If $f \in R^S$ is a regular R-valued ideal of S then the function $\bar{f} \in R^{S \times S}$ defined by $\bar{f}: (s, s') \mapsto f(s \ominus s')$ is an R-valued equivalence relation on S.

PROOF. If $s, s', s'' \in S$ then

$$\begin{aligned} \bar{f}(s, s')\bar{f}(s', s'') &= f(s \ominus s')f(s' \ominus s'') \preceq f([s \ominus s'] \oplus [s' \ominus s'']) \\ &= f(s \ominus s'') = \bar{f}(s, s'') \end{aligned}$$

and so $\bar{f}$ is transitive. Since f is regular, we have $\bar{f}(s, s) = f(0_S) = 1_R$ for all $s \in S$ and so $\bar{f}$ is reflexive. Finally, If $s, s' \in S$ then $\bar{f}(s, s') = f(s \ominus s') = f(s' \ominus s) = \bar{f}(s', s)$ and so $\bar{f}$ is symmetric. □

Therefore, in the situation of Propostion 8.12, for each $s \in S$ we have the equivalence class $\bar{f}_s \in R^S$ defined by $\bar{f}_s: s' \mapsto \bar{f}(s, s') = f(s \ominus s')$. By Proposition 2.6, we note that $\bar{f}_s = \bar{f}_{s'}$ if and only if $f(s \ominus s') = 1$. Set $S/f = \{\bar{f}_s \mid s \in S\}$ and define operations $\oplus$ and $\odot$ on S/f by setting $\bar{f}_s \oplus \bar{f}_{s'} = \bar{f}_{s \oplus s'}$; and $\bar{f}_s \odot \bar{f}_{s'} = \bar{f}_{s \odot s'}$. These operations are well-defined since:

(1) Assume that $\bar{f}_s = \bar{f}_t$ and $\bar{f}_{s'} = \bar{f}_{t'}$. If $s'' \in S$ then

$$\bar{f}_{s \oplus s'}(s'') = f(s \oplus s' \ominus s'')$$
$$= f([s \ominus t] \oplus [s' \ominus t'] \oplus [t \oplus t' \ominus s''])$$
$$\succeq f(s \ominus t)f(s' \ominus t')f(t \oplus t' \ominus s'')$$
$$= f(t \oplus t' \ominus s'')$$
$$= \bar{f}_{t \oplus t'}(s'')$$

and so $\bar{f}_{s \oplus s'} \succeq \bar{f}_{t \oplus t'}$. Similarly, $\bar{f}_{t \oplus t'} \succeq \bar{f}_{s \oplus s'}$ and so we have equality.

(2) Assume that $\bar{f}_s = \bar{f}_t$ and $\bar{f}_{s'} = \bar{f}_{t'}$. If $s'' \in S$ then

$$\bar{f}_{s \odot s'}(s'') = f(s \odot s' \ominus s'')$$
$$= f([s \odot s' \ominus s \odot t'] \oplus [s \odot t' \ominus t \odot t'] \oplus [t \odot t' \ominus s''])$$
$$\succeq f(s \odot [s' \ominus t'])f([s \ominus t] \odot t')f(t \odot t' \ominus s'')$$
$$\succeq f(s' \ominus t')f(s \ominus t)f(t \odot t' \ominus s'')$$
$$= f(t \odot t' \ominus s'')$$
$$= \bar{f}_{t \odot t'}(s'')$$

and so $\bar{f}_{s \odot s'} \succeq \bar{f}_{t \odot t'}$. Similarly, $\bar{f}_{t \odot t'} \succeq \bar{f}_{s \odot s'}$ and so we have equality.

It is easy to verify that, for any regular R-valued ideal f of S, the set S/f with the above operations is again a ring and that the function $\nu_f : s \mapsto \bar{f}_s$ is a surjective ring homomorphism from S onto S/f.

(8.13) PROPOSITION. Let R be a semiring satisfying $1 \succeq r$ for all $r \in R$ and let $(S, \oplus, \odot)$ be a ring. Let $f \in R^S$ be a regular R-valued ideal of S. Then

(1) $I = \{s \in S \mid f(s) = 1_R\}$ is an ideal of S.
(2) $S/f \cong S/I$.

PROOF. (1) If $s_1, s_2 \in I$ then $1_R \succeq f(s_1 \ominus s_2) \succeq f(s_1)f(-s_2) = f(s_1)f(s_2) = 1$ and so $s_1 \ominus s_2 \in I$. If $s \in S$ then $1_R \succeq f(ss_1) \succeq f(s_1) = 1_R$ and so $ss_1 \in I$. Similarly $s_1 s \in I$ and so I is an ideal of S.

(2) The map $\nu_f : s \mapsto \bar{f}_s$ is, as we have mentioned, a surjective ring homomorphism from S onto S/f. Moreover, $s \in ker(\nu_f) \Leftrightarrow \bar{f}_s = \bar{f}_0 \Leftrightarrow f(s \ominus 0) = 1_R \Leftrightarrow f(s) = 1_R \Leftrightarrow s \in I$, from which (2) follows immediately. $\square$

(8.14) PROPOSITION. Let R be a simple QLO-semiring. Let S and S' be rings and let $f \in R^S$ and $g \in R^{S'}$ be regular R-valued ideals of S and S' respectively. Let $\gamma : S \to S'$ be a ring homomorphism satisfying the condition that $h_\gamma[f] \preceq g$. Then there exists a ring homomorphism $\bar{\gamma} : S/f \to S'/g$ satisfying $\bar{\gamma}\nu_f = \nu_g \gamma$.

PROOF. Define $\bar{\gamma}$ by $\bar{\gamma} : \bar{f}_s \mapsto \bar{g}_{\gamma(s)}$. This function is well-defined and has the desired properties as a consequence of Proposition 8.11. $\square$

Let R be a semiring. An R-valued ideal f of a semiring $(S, \oplus, \odot)$ is *cancellative* if and only if $f(s) \succeq f(s \oplus s')f(s')$ and $f(s) \succeq f(s')f(s \oplus s')$ for all $s, s' \in S$. Thus, for example, if $R = (\mathbb{I}, \vee, \wedge)$ and if $(S, \oplus, \odot)$ is an arbitrary semiring then an R-valued ideal f of S is cancellative if and only if $\{s \in S \mid f(s) \geq a\}$ is a cancellative ideal of S for each $a \in \mathbb{I}$. In order to construct cancellative R-valued ideals, we need to extend to semirings a concept we defined earlier for semigroups. If $(S, \oplus, \odot)$ is a semiring then an R-valued equivalence relation $h \in R^{S \times S}$ is an *R-valued congruence* on S if and only if the following conditions are satisfied:

(1) $h(s_1 \oplus s_3, s_2 \oplus s_3) \succeq h(s_1, s_2)$;
(2) $h(s_1 \odot s_3, s_2 \oplus s_3) \succeq h(s_1, s_2)$; and
(3) $h(s_3 \odot s_1, s_3 \oplus s_2) \succeq h(s_1, s_2)$

for all $s_1, s_2, s_3 \in S$. Thus, for example, we have already noted that the function $h \in \mathbb{I}^{\mathbb{N} \times \mathbb{N}}$ defined by

$$h: (i,j) \mapsto \begin{cases} 1 & \text{if } i = j \\ \frac{1}{2} & \text{if } i + j \text{ is even} \\ 0 & \text{otherwise} \end{cases}$$

is an $\mathbb{I}$-valued equivalence relation on $\mathbb{N}$. In fact, it is an $\mathbb{I}$-valued congruence on $\mathbb{N}$.

(8.15) PROPOSITION. *Let R be a simple semiring an let $(S, \oplus, \odot)$ be a semiring. If $h \in R^{S \times S}$ is an R-valued congruence relation on S then the function $f: s \mapsto h(s, 0)$ is a cancellative regular R-valued ideal of S.*

PROOF. If $s, s' \in S$ then

$$f(s \oplus s') = h(s \oplus s', 0) \succeq h(s \oplus s', s')h(s', 0) \succeq h(s, 0)h(s', 0) = f(s)f(s').$$

Moreover,

$$f(s \odot s') = h(s \odot s', 0) = h(s \odot s', 0 \odot s') \succeq h(s, 0) = f(s)$$

and similarly $f(s' \odot s) \succeq f(s)$. Thus f is an R-valued ideal of S

$$f(s) = h(s, 0) \succeq h(s, s \oplus s')h(s \oplus s', 0) \succeq h(0, s')h(s \oplus s', 0)$$
$$= h(s', 0)h(s \oplus s', 0) = f(s')f(s \oplus s')$$

and similarly $f(s) \succeq f(s \oplus s')f(s')$. Thus f is cancellative. $\square$

Let $(S, \oplus, \odot)$ be a semiring and let R be an additively-idempotent semiring. If $f \in R^S$ is an R-valued ideal of S and if s is a unit in S then

$$f(1_S) = f(s \odot s^{-1}) \succeq f(s) \succeq f(1_S)$$

and so $f(s) = f(1_S)$. If the converse also holds then the R-valued ideal f of S is *quasilocal*.

(**8.16**) PROPOSITION. *Let $(S, \oplus, \odot)$ be a semiring and let R be an additively-idempotent semiring. If $f \in R^S$ is an R-valued ideal of S then f is quasilocal if and only if $H_f = \{s \in S \mid f(s) \succ f(1_S)\}$ is the unique maximal ideal of S.*

PROOF. We already know that H_f is an ideal of S. Note that f is quasilocal if and only if H_f is the set of all nonunits of S and that is true if and only if it is the unique maximal ideal of S. □

COROLLARY. *The following conditions on a semiring $(S, \oplus, \odot)$ are equivalent:*

(1) *The set of nonunits of S is an ideal;*
(2) *S has a quasilocal $\mathbb{B}$-valued ideal;*
(3) *There exist an additively-idempotent semiring R and a quasilocal R-valued ideal of S.*

PROOF. (1) ⇒ (2): If H is the ideal of S composed of all nonunits then the characteristic function $\chi_H \colon S \to \mathbb{B}$ is a quasilocal $\mathbb{B}$-valued ideal of S.

(2) ⇒ (3): This is immediate.

(3) ⇒ (1): This follows from Proposition 8.16. □

References

[1] K. S. Abdukhalikov, *The dual of a fuzzy subspace*, Fuzzy Sets and Systems **82** (1995), 375 – 381.

[2] _____, M. S. Tulenbaev & U. U. Umirbaev, *On fuzzy bases of vector spaces*, Fuzzy Sets and Systems **63** (1994), 201 – 206.

[3] M. T. Abu Osman, *On some product of fuzzy subgroups*, Fuzzy Sets and Systems **24** (1987), 79 – 86.

[4] Salah Abu-Zaid, *On fuzzy subgroups*, Fuzzy Sets and Systems **55** (1993), 237 – 240.

[5] A. V. Aho, J. E. Hopcraft & J. D. Ullman, *The Design and Analysis of Computer Algorithms*, Addison-Wesley, Reading, MA, 1974.

[6] J. Ahsan, M. Farid Khan, & M. Shabir, *Characterizations of monoids by the properties of their fuzzy subsystems*, Fuzzy Sets and Systems **56** (1993), 199 – 208.

[7] J. Ahsan, K. Saifullah & M. Farid Khan, *Fuzzy semirings*, Fuzzy Sets and Systems **60** (1993), 309 – 320.

[8] Naseem Ajmal, *Homomorphism of fuzzy groups, correspondence theorem and fuzzy quotient groups*, Fuzzy Sets and Systems **61** (1994), 329 – 339.

[9] _____, *Fuzzy groups with sup property*, Info. Sci. **93** (1996), 247 – 264.

[10] _____ & K. V. Thomas, *The lattices of fuzzy subgroups and fuzzy normal subgroups*, Info. Sci. **76** (1994), 1 – 11.

[11] _____ & K. V. Thomas, *A complete study of the lattices of fuzzy congruences and fuzzy normal subgroups*, Info. Sci. **82** (1995), 197 – 218.

[12] Mustafa Akgül, *Some properties of fuzzy groups*, J. Math. Anal. Appl. **133** (1988), 93 – 100.

[13] Marianne Akian, *Densities of idempotent measures and large deviations*, (preprint, 1995).

[14] _____, *Theory of cost measures: convergence of decision variables*, (preprint, 1995).

[15] _____, Jean-Pierre Quadrat & Michel Viot, *Bellman processes*, 11th International Conference on Analysis and Optimization of Systems: Discrete Event Systems (G. Cohen & J.-P. Quadrat, eds.), Springer-Verlag, Berlin, 1994.

[16] _____, Jean-Pierre Quadrat & Michel Viot, *Duality between probability and optimization*, Idempotency (Bristol 1994) (J. Gunawardena, ed.), Publ. Newton Inst. #11, Cambridge Univ. Press, Cambridge, 1998, pp. 331 – 353.

[17] Francisco E. Alarcón & D. D. Anderson, *Commutative semirings and their lattices of ideals*, Houston J. Math. **20** (1994), 571 – 590.

[18] Mustafa A. Amer & Nehad A. Morsi, *Bounded linear transformations between probabilistic normed vector spaces*, Fuzzy Sets and Systems **73** (1995), 167 – 183.

[19] Charles André, *An algebra for "SENS"*, Technical Report LASSY-13S RR 89-8, Laboratoire de Signaux et Systèmes, Université de Nice-Sophia Antipolis / CNRS, Nice, 1989.

[20] J. M. Anthony & Howard Sherwood, *Fuzzy groups redefined*, J. Math. Anal. Appl. **69** (1979), 124 – 130.

[21] ———, *A characterization of fuzzy subgroups*, Fuzzy Sets and systems **7** (1982), 297 – 305.

[22] Johannes Arz, *Syntactic congruences and syntactic algebras*, RAIRO Informatique Théorique **17** (1983), 231 – 238.

[23] M. Asaad & Salah Abou-Zaid, *Characterization of fuzzy subgroups*, Fuzzy Sets and Systems **63** (1996), 247 – 251.

[24] ———, *A contribution to the theory of fuzzy subgroups*, Fuzzy Sets and Systems **77** (1996), 355 - 369.

[25] Jonathan Babb, Matthew Frank & Anant Agarwal, *Solving graph problems with dynamic computation structures*, Proceedings of the SPIE Photonics-East Symposium: Reconfigurable Technology for Rapid Product Development and Computing, Boston, 1996.

[26] F. Baccelli, G. Cohen, G. J. Olsder & J. - P. Quadrat, *Synchronization and Linearity*, John Wiley & Sons, New York, 1992.

[27] Hans Bandemer & Siegfried Gottwald, *Fuzzy Sets, Fuzzy Logic, Fuzzy Methods*, John Wiley & Sons, New York, 1996.

[28] W. Bandler & L. J. Kohout, *Fuzzy relational products as a tool for analysis and synthesis of the behaviour of complex natural and artificial systems*, Fuzzy Sets: Theory and Application to Policy Analysis and Information Systems (P. P. Wang & S. K. Chang, eds.), Plenum, New York, 1980, pp. 341 – 367.

[29] ———, *A survey of fuzzy relational products in their applicability to medicine and clinical psychology*, Knowledge Representation in Medicine and Clinical Behavioural Sciences (W. Bandler & L. J. Kohout, eds.), Abacus Press, Cambridge, 1986, pp. 107 – 118.

[30] ———, *Semantics of implication operators and fuzzy relational products*, Int. J. Man-Machine Studies **12** (1980), 89 – 116.

[31] Twan Basten, *Parsing partially ordered multisets*, Int. J. Found. Comp. Sci,. **8** (1997), 379 – 407.

[32] C. Benzaken, *Structures algébriques des cheminements: pseudo-trellis, gerbiers de carré nul*, Network and Switching Theory (G. Biorci, ed.), Academic Press, New York, 1968, pp. 40 – 47.

[33] Rudolph Berghammer, Peter Kempf, Gunther Schmidt & Thomas Ströhlein, *Relation algebras and the logic of programs*, Algebraic Logic (H. Andréka et al., eds.), Colloq. Math. Soc. János Bolyai #54, North Holland, Amsterdam, 1991.

[34] Noël Bérnard, *Neighbors, elements, graduation: the multiensembles*, The Mathematics of Fuzzy Systems (Antonio Di Nola & Aldo Ventre, eds.), Verlag TÜV Rheinland, Köln, 1986, pp. 1 – 31.

[35] Ugo Berni-Canani. Francis Borceux & Rosanna Succi-Cruciani, *A theory of quantal sets*, J. Pure Appl. Algebra **62** (1989), 123 – 136.

[36] S. K. Bhakat & P. Das, *On the definition of a fuzzy subgroup*, Fuzzy Sets and Systems **51** (1992), 235 – 241.

[37] P. Bhattacharya & N. P. Mukherjee, *Fuzzy relations and fuzzy groups*, Info. Sci. **36** (1985), 267 – 282.

[38] Garrett Birkhoff, *Moore-Smith convergence in general topology*, Ann. Math. (2) **38** (1937), 39 – 56.

[39] _____, *Lattice Theory*, American Mathematical Society, Providence, 1940.

[40] Stefano Bistarelli, Ugo Montanari & Francesca Rossi, *Semiring-based constraint satisfaction and optimization*, (preprint, 1997).

[41] Ranjit Biswas, *Rosenfeld's fuzzy subgroups with interval-valued membership functions*, Fuzzy Sets and Systems 63 (1994), 87 – 90.

[42] Wayne D. Blizard, *Multiset theory*, Notre Dame J. Formal Logic **30** (1989), 36 – 66.

[43] _____, *Real-valued multisets and fuzzy sets*, Fuzzy Sets and Systems **33** (1989), 77 – 97.

[44] Stephen E. Bloom & Zoltán Ésik, *Free shuffle algebras in language varieties*, Theor. Comp. Sci, **163** (1996), 55 – 98.

[45] T. S. Blyth & M. F. Janowitz, *Residuation Theory*, Pergamon, Oxford, 1972.

[46] M. Boffa, *Une condition impliquant toutes des identités rationnelles*, RAIRO Inform. Théor. Appl. **29** (1995), 515 – 518.

[47] Kenneth P. Bogart, *Intervals and orders: what comes after interval orders?*, Orders, Algorithms and Applications (Vincent Boichitté & Michel Morvan, eds.), Lecture Notes in Computer Science #831, Springer-Verlag, Berlin, 1994, pp. 13 – 32.

[48] Nicholas Bourbaki, *General Topology, part 2*, Hermann, Paris, 1966.

[49] Mary M. Bourke & D. Grant Fisher, *The complete resolution of Cartesian products of fuzzy sets*, Fuzzy Sets and Systems **63** (1994), 111 – 115.

[50] _____, *Solution algorithms for fuzzy relational equations with max-product composition*, Fuzzy Sets and Systems **94** (1998), 61 – 69.

[51] J. G. Braker & G. J. Olsder, *The power algorithm in max algebra*, Linear Algebra and its Appl. **182** (1993), 67 – 89.

[52] Chris Brink, *Power structures*, Algebra Universalis **30** (1993), 177 – 216.

[53] G. Burosch, J. Demetroics, & G. O. H. Katona, *The poset of closures as a model of changing databases*, Order **4** (1987), 127 – 142.

[54] Michael Bussiek, Hannes Hassler, Gerhard J. Woeginger & Uwe T. Zimmermann, *Fast algorithms for the maximal convolution problem*, Operations Research Leters **15** (1994), 133 – 141.

[55] Dan Butnariu & Erich P. Klement, *Triangular Norm-Based Measures and Games with Fuzzy Coalitions*, Kluwer, Dordrecht, 1993.

[56] _____, Eric P. Klement & Samy Zafrany, *On triangular norm-based propositional fuzzy logics*, Fuzzy Sets and Systems **69** (1995), 241 – 255.

[57] Cao Zhiqiang, *An algebraic system generalizing the fuzzy subsets of a set*, Advances in Fuzzy Sets, Possibility Theory, and Applications (Paul Wang, ed.), Plenum, New York, 1993, pp. 71 – 80.

[58] _____, Ki Hang Kim, & Fred W. Roush, *Incline Algebra and its Applications*, Ellis Horwood, Chichester, 1984.

[59] Bernard A. Carré, *Graphs and Networks*, Oxford Univ. Press, Oxford, 1979.

[60] A. B. Chakraborty & S. S. Khare, *Fuzzy homomorphism and algebraic structures*, Fuzzy Sets and Systems **59** (1993), 211 – 221.

[61] C. C. Chang, *Algebraic analysis of many valued logics*, Trans. Amer. Math. Soc. **88** (1958), 467 – 490.

[62] Chen Lichun & Peng Boxing, *The fuzzy relation equation with union or intersection preserving operator*, Fuzzy Sets and Systems **25** (1988), 191 – 204.

[63] B. P. Chisala & M.-M. Mawanda, *Counting measure for Kuratowski finite parts and decidability*, Cahier de Topologie et Geométrie Différential Catégoriques **23** (1991), 345 – 353.

[64] Kenneth L. Clarkson, *Nearest neighbor queries in metric spaces*, Proceedings of the 29th Annual ACM Symposium on Theory of Complexity, Association for Computing Machinery, New York, 1997, pp. 609 – 617.

[65] E. F. Codd, *A relational model of data for large shared data banks*, Comm. ACM **13** (1970), 377 – 387.

[66] John H. Conway, *Regular Algebra and Finite Machines*, Chapman and Hill, London, 1971.

[67] J. Coulon, J.-H. Coulon & U Höhle, *Classification of extremal subobjects of algebras over* **SM-SET**, Applications of Category theory to Fuzzy Subsets (S. E. Rodabaugh et al., eds.), Kluwer, Dordrecht, 1992, pp. 9 – 31.

[68] Raymond A. Cuninghame-Green, *Describing industrial processes with interference and approximating their steady-state behavior*, Operational Research Quarterly **13** (1962), 95 – 100.

[69] _____, *Minimax Algebra*, Lecture Notes in Economics and Mathematical Systems #166, Springer-Verlag, Berlin, 1979.

[70] R. A. Cuninghame-Green & Katarína Cechlárová, *Residuation in fuzzy algebra and some applications*, Fuzzy Sets and Systems **71** (1995), 227 – 239.

[71] E. Damiani, O. D'Antona & D. Loeb, *Getting results with negative thinking*, Actes de l'atelier de Combinatoire Franco-Quebois, Bordeaux, May 6-7, 1997 (G. Labelle & J.-G. Penaud, eds.), Publications du LACIM, Montreal, 1992, pp. 191 – 214.

[72] G. B. Dantzig, W. D. Blattner & M. R. Rao, *Finding a cycle in a graph with minimum cost to time ratio with application to a ship routing problem*, Théorie des Graphes, Proc. of the Int. Symp. Rome, Italy, Dunod, Paris, 1967.

[73] Phullendu Das, *Fuzzy vector spaces under triangular norms*, Fuzzy Sets and Systems **25** (1988), 73 – 85.

[74] Bernard De Baets & E. E. Kerre, *The generalized modus ponens and triangular fuzzy data model*, Fuzzy Sets and Systems **59** (1993), 305 – 317.

[75] _____, *Fuzzy relational compositions*, Fuzzy Sets and Systems **60** (1993), 109 – 120.

[76] Bernard De Baets & Andrea Marková-Stupňanová, *Analytical expressions for the addition of fuzzy intervals,*, Fuzzy Sets and Systems **91** (1997), 203 – 213.

[77] Pierre Del Moral, *Maslov optimization theory, optimality vs. randomness*, appended to Vasilli N. Kolokol'tsov & Victor P. Maslov: *Idempotent Analysis and Its Applications* (1997), Kluwer, Dordrecht.

[78] A. P. Dempster, *Upper and lower probabilities induced by a multivaluated mapping*, Ann. Math. Statist. **38** (1967), 325 – 339.

[79] M.-M. Deza & I. G. Rosenberg, *General convolutions motivated by designs*, Acta Univ. Carolinae Math. et Phys. **27** (1986), 49 – 66.

[80] Antonio Di Nola, Witold Pedrycz, & Salvatore Sessa, *Fuzzy relation equations with equality and difference composition operators*, Fuzzy Sets and Systems **25** (1988), 205 – 215.

[81] Antonio Di Nola, Witold Pedrycz, & Salvatore Sessa, *Fuzzy relational structures: the state of the art*, Fuzzy Sets and Systems **75** (1995), 241 – 262.

[82] Antonio Di Nola, Witold Pedrycz, Salvatore Sessa, & W. Pei-Zhuang, *Fuzzy relation equations under a class of triangular norms: a survey and new results*, Stochastica **2** (1984), 99 – 145.

[83] Antonio Di Nola, Salvatore Sessa & Witold Pedrycz, *Decomposition problem of fuzzy relations*, Inter. J. General Systems **10** (1985), 123 – 133.

[84] Antonio Di Nola, Salvatore Sessa & Witold Pedrycz, *On some finite fuzzy relaxation equations*, Info. Sci. **50** (1990), 93 – 109.

[85] V. N. Dixit, Rajesh Kumar & Naseem Adjmal, *Level subgroups and union of fuzzy subgroups*, Fuzzy Sets and Systems **37** (1990), 350 – 371.

[86] J.-P. Doignon, B. Monjaret, M. Roubens & Ph. Vincke, *Biorders families, valued relations and preference modelling*, J. Math. Psych. **30** (1986), 435 – 480.

[87] Didier Dubois & Henri Prade, *Fuzzy Sets and Systems, Theory and Applications*, Academic Press, New York, 1980.

[88] _____, *Fuzzy numbers, an overview*, Analysis of Fuzzy Information, (J. C. Bezdek, ed.), vol. 2, CRC Press, Boca Raton, 1988, pp. 3 – 39.

[89] _____, *Toll sets*, Proceedings of IFSA'91, Brussels, 1991, pp. 21 – 24.

[90] _____ *Fuzzy relation equations and causal reasoning*, Fuzzy Sets and Systems **75** (1995), 119 – 134.

[91] M. L. Dubreil-Jacotin, L. Lesieur, & R. Croisot, *Leçons sur la théorie des treillis, des structures algébriques ordonées et des treillis géometriques*, Gauthier-Villars, Paris, 1953.

[92] P. I. Dudnikov & S. N. Samborskiĭ, *Endomorphisms of semimodules over semirings with an idempotent operation*, Izv. Akad. Nauk SSSR Ser. Mat. **55** (1991), 93 – 103.

[93] _____, *Endomorphisms of finitely generated free semimodules*, Idempotent Analysis (V. P. Maslov & S. N. Samborskiĭ, eds.), Advances in Soviet Mathematics #13, American Mathematical Society, Providence, 1992, pp. 65–85.

[94] Florence Dupin de Saint-Cyr, Jérôme Lang & Thomas Schiex, *Penalty logic and its link with Dempster-Shafer theory*, Proceedings of the 10th Conference on Uncertainty in Artificial Intelligence (R. Lopez de Mantaras & D. Poole, eds.), Morgan Kaufmann, 1994.

[95] T. K. Dutta & B. K. Biswas, *Fuzzy prime ideals of a semiring*, Bull. Malaysian Math. Soc. (Second Ser.) **17** (1994), 9 – 16.

[96] _____, *Fuzzy congruence and quotient semiring of a semiring*, J. Fuzzy Math. **4** (1996), 737 – 748.

[97] _____, *Structure of fuzzy ideals of semirings*, Bull. Cal. Math. Soc. **89** (1997), 271 – 284.

[98] Samuel Eilenberg, *Automata, Languages and Machines*, vol. A, Academic Press, New York, 1974.

[99] M. A. Erceg, *Functions, equivalence relations, quotient spaces, and subsets in fuzzy set theory*, Fuzzy Sets and Systems **3** (1980), 75 – 92.

[100] M. S. Eroğlu, *The homomorphic image of a fuzzy subgroup is always a fuzzy subgroup*, Fuzzy Sets and Systems **33** (1989), 255 – 256.

[101] Esfandiar Eslami & John N. Mordeson, *Structure of fuzzy subrings*, Info. Sci. **76** (1994), 57 – 65.

[102] Fang Jin-xuan, *Fuzzy homomorphism and fuzzy isomorphism*, Fuzzy Sets and Systems **63** (1994), 237 – 242.

[103] László Filep, *Structure and construction of fuzzy subgroups of a group*, Fuzzy Sets and Systems **51** (1992), 105 – 109.

[104] _____, *Studies in fuzzy relations using triangular norms*, Info. Sci. **67** (1993), 127 – 135.

[105] _____, *Structure of L-fuzzy groups and relations*, J. Fuzzy Math. **2** (1994), 871 – 892.

[106] A. V. Finkelstein & M. A. Roytberg, *Computation of biopolymers: a general approach to different problems*, BioSystems **30** (1993), 1 – 20.

[107] Isidore Fleischer, *Maximality and ultracompleteness in normed modules*, Proc. Amer. Math. Soc. **9** (1958), 151 – 157.

[108] W. M. Fleischman (ed.), *Set-Valued Mappings, Selections and the Topological Properties of 2^X*, Lecture Notes in Mathematics #171, Springer-Verlag, Berlin, 1970.

[109] János C. Fodor, *On fuzzy implication operators*, Fuzzy Sets and Systems **42** (1991), 293 – 300.

[110] _____, *Traces of fuzzy binary relations*, Fuzzy Sets and Systems **50** (1992), 331 – 341.

[111] Michael L. Fredman, *Arithmetical convolution products and generalizations*, Duke Math. J. **37** (1970), 231 – 242.

[112] Orrin Frink, *Topology in lattices*, Trans. Amer. Math. Soc. **51** (1942), 569 – 582.

[113] R. Fuller & H.-J. Zimmermann, *On computation of the compositional rule of inference under triangular norms*, Fuzzy Sets and Systems **51** (1992), 267 – 275.

[114] Peter Gabriel, *Indecomposable representations I*, Manuscripta Math. **6** (1972), 71 – 103.

[115] Siegfried Gähler & Werner Gähler, *Fuzzy real numbers*, Fuzzy Sets and Systems **66** (1994), 137 – 158.

[116] Werner Gähler, *Fuzzy topology*, Mathematical Research 66, Akademie Verlag, Berlin, 1992.

[117] Barry J. Gardner, *Some abstract algebra from the elementary calculus course*, Int. J. Math. Educ. Sci. Technol. **24** (1993), 781-789.

[118] Max Garzon & Godfrey C. Muganda, *Free fuzzy groups and fuzzy group presentations*, Fuzzy Sets and Systems **48** (1992), 249 – 255.

[119] Stéphane Gaubert, *Systèmes Dynamiques à Événements Discrets*, (Notes de cours commun, ENSMP, Option Automatique & DEA ATS Orsay), INRIA Rocquencourt, Le Chesnay, 1996.

[120] _____, *Two lectures on max-plus algebra*, Proceedings of the 26th Spring School of Theoretical Informatics (INRIA 1998), INRIA Rocquencourt, Le Chesnay, 1998, pp. 81 – 146.

[121] _____ & J. Mairesse, *Task resource models and $(max,+)$ automata*, Idempotency (Bristol 1994) (J. Gunawardena, ed.), Publ. Newton Inst. #11, Cambridge Univ. Press, Cambridge, 1998, pp. 133 – 144.

[122] _____ & Max Plus, *Methods and applications of $(max,+)$ linear algebra*, STACS 97 (Rüdiger Reischuk & Michel Morvan, eds.), Lecture Notes in Computer Science #1200, Springer-Verlag, Berlin, 1997.

REFERENCES

[123] Yan Georget & Philippe Codognet, *Compiling semiring-based constraints with* clp(FD,S), submitted for publication (1998).

[124] Giangiacomo Gerla, *Pavelka's fuzzy logic and free L-subsemigroups*, Zeitschr. f. math. Logik und Grundlagen d. Math. **31** (1985), 123 – 129.

[125] _____, *On the concept of fuzzy point*, Fuzzy Sets and Systems **18** (1986), 159 – 172.

[126] _____, *Generalized fuzzy points*, J. Math. Anal. Appl. **120** (1986), 761 – 168.

[127] _____, *Distances, diameters and versimilitude of theories*, Arch Math. Logic **31** (1992), 407 – 414.

[128] M. H. Ghanim, *L-Fuzzy basic proximity spaces*, Fuzzy Sets and Systems **27** (1988), 197 – 203.

[129] Phan H. Giang, *Representation of uncertan belief using interval probability*, Proceedings of the 27th International Symposium on Multiple-Valued Logic, IEEE Computer Society Press, 1997, pp. 111 – 116.

[130] C. R. Giardina & E. R. Dougherty, *Morphological Methods in Image and Signal Processing*, Prentice Hall, NJ, 1988.

[131] G. Gierz, K. H. Hoffman, K. Keimel, J. D. Lawson, M. Mislove & D. S. Scott, *A Compendium of Continuous Lattices*, Springer-Verlag, Berlin, 1980.

[132] B. Giffler, *Scheduling general production systems using schedule algebra*, Naval Res. Logist. Quart. **10** (1963), 237 – 255.

[133] _____, *Schedule algebra: a progress report*, Naval Res. Logist. Quart. **15** (1968), 255 - 280.

[134] Robert Gilmer, *Multiplicative Ideal Theory*, Marcel Dekker, New York, 1972.

[135] _____, *Commutative Semigroup Rings*, University of Chicago Press, Chicago, 1984.

[136] Jean-Yves Girard, *Linear logic: its syntax and semantics*, Advances in Linear Logic, London Math. Soc. Lecture Notes # 222, Cambridge University Press, Cambridge, 1995.

[137] Jay L. Gischer, *The equational theory of pomsets*, Theor. Comp. Sci. **61** (1988), 199–224.

[138] J. A. Goguen, *L-fuzzy sets*, J. Math. Anal. Appl. **18** (1967), 145–174.

[139] _____, *The logic of inexact concepts*, Synthese **19** (1969), 325 – 373.

[140] Jonathan S. Golan, *Decomposition and Dimension in Module Categories*, Lecture Notes in Pure and Applied Mathematics #33, Marcel Dekker, New York, 1977.

[141] _____, *Torsion Theories*, Longman Scientific & Technical, Harlow, 1986.

[142] _____, *Linear Topologies on a Ring: an Overview*, Longman Scientific & Technical, Harlow, 1987.

[143] _____, *Making modules fuzzy*, Fuzzy Sets and Systems **32** (1989), 91 – 94.

[144] _____, *More topologies on the torsion-theoretic spectrum of a ring*, Periodica Math. Hungar. **21** (1990), 257 – 260.

[145] _____, *Information semimodules and absorbing subsemimodules*, Journal of Mathematics (Hanoi) **19** (1991), 1 – 21.

[146] _____, *The Theory of Semirings, with Applications in Mathematics and Theoretical Computer Science*, Longman Scientific & Technical, Harlow, 1992.

[147] _____, *Dijkstra semirings and their use in characterizing fuzzy and toll connectives*, Fuzzy Systems and A. I. **3** (1994), 3 – 14.

[148] _____, *Norms, semirings, and power algebras*, Proceedings of the Fourth Ramanujan Symposium on Algebra and its Applications (S. Parvathi et al., eds.), Ramanujan Institute for Advanced Study in Mathematics, Madras, 1996.

[149] _____, *Semirings and Their Applications*, Kluwer, Dordrecht, 1999.

[150] _____, *Semiring-valued quasimetrics on the set of submodles of a module*, to appear, Math. J. Okayama Univ. (1999).

[151] _____ & Harold Simmons, *Derivatives, Nuclei and Dimensions on the Frame of Torsion Theories*, Longman Scientific & Technical, Harlow, 1988.

[152] _____ & Huaxiong Wang, *On embedding in complete semirings*, Comm Algebra **24** (1996), 2945–2962.

[153] Martin Goldstern, *Vervellständigung von Halbringen*, Diplomarbeit, T.U. Wien, 1986.

[154] Michel Gondran & Michel Minoux, *Graphs and Algorithms*, Wiley-Interscience, New York, 1984.

[155] _____, *Linear algebra in dioïds: a survey of recent results*, Ann. Discrete Math. **19** (1984), 147 – 164.

[156] León González, *A note on the infinitary action of triangular norms and conorms*, Fuzzy Sets and Systems **101** (1999), 177 – 180.

[157] _____ & Angel Marin, *Weak properties and aggregated extension of fuzzy relations*, Fuzzy Sets and Systems **85** (1997), 311 – 318.

[158] Siegfried Gottwald, *Fuzzy set theory with t-norms and φ-operators*, The Mathematics of Fuzzy Systems (Antonio Di Nola & Aldo Ventre, eds.), Verlag TÜV Rheinland, Köln, 1986, pp. 143 – 195.

[159] J. Grabowski, *On partial languages*, Fund. Informat. **4** (1981), 427 – 98.

[160] J. E. Graver & W. B. Jurkat, *Algebra structures of general design*, J. Alg. **23** (1972), 574 – 589.

[161] Wen-Xiang Gu & De-Gang Chen, *A fuzzy subroupoid which is not a fuzzy group*, Fuzzy Sets and Systems **62** (1994), 115 – 116.

[162] Wen-Xiang Gu, Su-Yun Li, De-Gang Chen, & Yao-Hua Lu, *The generalized t-norms and the TLPF-groups*, Fuzzy Sets and Systems **72** (1995), 357–364.

[163] Jeremy Gunawardena, *Cyclic times and fixed points of min-max functions*, 11th International Conference on Analysis and Optimization of Systems (G. Cohen & J. P. Quadrat, eds.), Lecture Notes in Computer Science #199, Springer Verlag, Berlin, 1994, pp. 266 – 273.

[164] _____, *Min-Max functions*, Discrete Event Dynamical Systems **4** (1994), 377 – 406.

[165] _____, *An introduction to idempotency*, Idempotency (Bristol 1994) (J. Gunawardena, ed.), Publ. Newton Inst. #11, Cambridge Univ. Press, Cambridge, 1998, pp. 1 – 49.

[166] K. C. Gupta & R. K. Gupta, *Fuzzy equivalence relation redefined*, Fuzzy Sets and Systems **79** (1996), 227 – 233.

[167] K. C. Gupta & Manoj K. Kantroo, *The intrinsic product of fuzzy subsets of a ring*, Fuzzy Sets and Systems **57** (1993), 103 – 110.

[168] K. C. Gupta & Suryansu Ray, *Modularity of the quasi-hamiltonian fuzzy subgroups*, Info. Sci. **79** (1994), 233 – 250.

[169] M. Gupta & T. Yamakawa, *Fuzzy Computing, Theory, Hardware, and Applications*, Elsevier, Amsterdam, 1988.

[170] Vineet Gupta, *Chu Spaces: a Model of Concurrency*, PhD. Dissertation, Department of Computer Science, Stanford University, 1994.

[171] Peter Hamburg, *Fuzzy sets and De Morgan algebras*, Fuzzy Sets and Systems **27** (1988), 21 – 29.

[172] Tom Head, *A metatheorem for deriving fuzzy theorems from crisp versions*, Fuzzy Sets and Systems **73** (1995), 349 – 358.

[173] Udo Hebisch, *Eine algebraische Theorie unendlicher Summen mit Anwendungen auf Halbgruppen und Halbringe*, Bayreuther Math. Schriften **40** (1992), 21 – 152.
[174] _____ & Hanns J. Weinert, *Semirings without zero divisors*, Math. Pannon. **1** (1990), 73 – 94.
[175] _____ & Hanns J. Weinert, *Halbringe*, Teubner, Stuttgart, 1993.
[176] _____ & Hanns J. Weinert, *Semirings and semifields*, Handbook of Algebra (M. Hazewinkel, ed.), vol. I, Elsevier, Amsterdam, 1966.
[177] Reinhold Heckmann, *Power domain constructions*, Sci. Comp. Program. **17** (1991), 77 – 117.
[178] _____, *Observable modules and power domain constructions*, Algebra, Logic, and Applications **5** (1993), 159 – 187.
[179] _____, *Power domains and second-order predicates*, Theoretical Comp. Sci. **111** (1993), 59 – 88.
[180] H. Hellendoorn, *Closure properties of the compositional rule of inference*, Fuzzy Sets and Systems **35** (1990), 163 – 183.
[181] S. Heilpern, *Representation and application of fuzzy numbers*, Fuzzy Sets and Systems **91** (1997), 259 – 268.
[182] José A. Herencia, *Graded sets and points: A stratified approach to fuzzy sets and points*, Fuzzy Sets and Systems **77** (1996), 191 – 202.
[183] Wim H. Hesselink, *Command algebras, recursion and program transformation*, Formal Aspects of Computing **2** (1990), 60 – 104.
[184] Einar Hille & Ralph S. Phillips, *Functional Analysis and Semi-groups*, AMS Colloquium Publication XXXI, revised edition, American Mathematical Society, Providence, 1957.
[185] Ulrich Höhle & Alexander Šostak, *A general theory of fuzzy topological spaces*, Fuzzy Sets and Systems **73** (1995), 131 – 149.
[186] Wladyslaw Homenda & Witold Pedrycz, *Processing uncertain information in the linear space of fuzzy sets*, Fuzzy Sets and Systems **44** (1991), 187 – 198.
[187] Yuang-Cheh Hsueh, *Extended lattice-ordered structures for L-fuzzy sets and L-fuzzy numbers*, Fuzzy Sets and Systems **54** (1993), 81 – 90.
[188] K. Iseki, *A characterization of regular semigroups*, Proc. Japan Acad. **32** (1965), 676 – 677.
[189] J. Jacas & L. Valverde, *On fuzzy relations, metrics, and cluster analysis*, (preprint, 1996).
[190] K. Jensen, *Coloured Petri Nets, vol 1*, Springer-Verlag, Berlin, 1992.
[191] _____, *Coloured Petri Nets, vol 2*, Springer-Verlag, Berlin, 1995.
[192] Peter T. Johnstone, *Stone Spaces*, Cambridge University Press, Cambridge, 1982.
[193] Milan V. Jovanović and Veselin M. Jungić, *Algebraic set operations, multifunctions, and indefinite integrals*, Mathematics Magazine **69** (1996), 350 – 354.
[194] Young Bae Jun, J. Neggers & Hee Sik Kim, *Normal L-fuzzy ideals in semirings*, Fuzzy Sets and Systems **82** (1996), 383 – 386.
[195] Georg Karner, *On limits in complete semirings*, Semigroup Forum **45** (1992), 148 – 165.
[196] _____, *A topology for complete semirings*, Proceedings of STACS '94 (P. Enjalbert et al., eds.), Lecture Notes in Computer Science # 775, Springer-Verlag, Berlin, pp. 389 – 400.
[197] A. K. Katsaras & D. B. Liu, *Fuzzy vector spaces and fuzzy topological vector spaces*, J. Math. Anal. Appl. **58** (1977), 135 – 146.

[198] Arnold Kaufmann & Madan M. Gupta, *Introduction to Fuzzy Arithmetic*, Van Nostrand Reinhold, New York, 1985.

[199] Chang Bum Kim & Mi-Ae Park, *k-Fuzzy ideals in semirings*, Fuzzy Sets and Systems **81** (1996), 281 – 286.

[200] J. B. Kim, Y. H. Kim, & C. B. Kim, *Resolution of the Cartesian product of fuzzy sets*, Fuzzy Sets and Systems **41** (1991), 293 – 297.

[201] Jae-Gyeom Kim, *On groups and fuzzy subgroups*, Fuzzy Sets and Systems **67** (1994), 347-348.

[202] _____, *Commutative fuzzy sets and nilpotent fuzzy groups*, Info. Sci. **83** (1995), 161 – 174.

[203] _____, *Lattices of fuzzy subgroupoids, fuzzy submonoids, and fuzzy subgroups*, Info. Sci. **91** (1996), 77 – 93.

[294] _____, *Meet-irreducibility of fuzzy subgroups*, Fuzzy Sets and Systems **91** (1997), 389 – 397.

[205] Ju Pil Kim & Deok Rak Bae, *Fuzzy congruences in groups*, Fuzzy Sets and Systems **85** (1997), 115 – 120.

[206] K. H. Kim & F. W. Roush, *Inclines of algebraic structures*, Fuzzy Sets and Systems **72** (1995), 189 – 196.

[207] Stephen C. Kleene, *Representation of events in nerve nets and finite automata*, Automata Studies (C. E. Shannon & J. McCarthy, eds.), Princeton University Press, Princeton, 1956.

[208] G. Klir & T. A. Folger, *Fuzzy Sets, Uncertainty, and Information*, Prentice Hall, Englewood Cliffs, N. J., 1988.

[209] Donald Knuth, *The Art of Computer Programming, vol. 2: Seminumerical algorithms*, Addison-Wesley, Reading, Mass., 1969.

[210] Nami Kobayashi, *The closure under division and a characterization of the recognizable Z-subsets*, RAIRO Inform. Théor. Appl. **30** (1996), 209 – 230.

[211] Vassili N. Kolokol'tsov, *On linear, additive, and homogeneous operators in idempotent analysis*, Idempotent Analysis (V. P. Maslov & S. N. Samborskiĭ, eds.), Advances in Soviet Mathematics #13, American Mathematical Society, Providence, 1992.

[212] Vassili N. Kolokol'tsov & Victor P. Maslov, *The general form of the endomorphisms in the space of continuous functions with values in a numerical commutative semiring*, Soviet Math. Dokl. **36** (1988), 55 – 59.

[213] _____, *Idempotent Analysis and Its Applications*, Kluwer, Dordrecht, 1997.

[214] Ralph Kopperman, *All topologies come from generalized metrics*, Amer. Math. Monthly **95** (1988), 89 – 97.

[215] B. Korte, L. Lovász & R. Schrader, *Greedoids*, Springer-Verlag, Berlin, 1991.

[216] Maurice Kostas, *Groupoïdes, demihypergroupes, et hypergroupes*, J. Math Pures et Appl. **49** (1970), 155 – 192.

[217] Daniel Krob, *Monoids et semi-anneaux complets*, Semigroup Forum **36** (1987), 323 – 339.

[218] Salma Kuhlmann, *Valuation bases for extensions of valued vector spaces*, Forum Math **8** (1996), 723 – 735.

[219] W. Kuich & F. J. Urbanek, *Infinite linear systems and one counter languages*, Theor. Comp. Sci. **22** (1983), 95 – 126.

[220] I. J. Kumar, P. K. Saxena & Pratibha Yadav, *Fuzzy normal subgroups and fuzzy quotients*, Fuzzy Sets and Systems **46** (1992), 121 – 132.

[221] Rajesh Kumar, *Fuzzy vector spaces and fuzzy cosets*, Fuzzy Sets and Systems **45** (1992), 109 – 116.

[222] _____, *Fuzzy subgroups, fuzzy ideals, and fuzzy cosets: some properties*, Fuzzy Sets and Systems **48** (1992), 121 – 132.

[223] _____, *On the dimension of a fuzzy subspace*, Fuzzy Sets and Systems **54** (1993), 229 – 234.

[224] H. V. Kumbhojkar & M. S. Bapat, *Not-so-fuzzy fuzzy ideals*, Fuzzy Sets and Systems **37** (1990), 237 – 243.

[225] J. Kuntzmann, *Theorie des Reseaux (Graphes)*, Dunod, Paris, 1972.

[226] Takashi Kuraoka & Nobuaki Kuroki, *On fuzzy quotient rings induced by fuzzy ideals*, Fuzzy Sets and Systems **47** (1992), 381 – 386.

[227] Takaskhi Kuraoka & Nobu-Yuki Suzuki, *A simple characterization of fuzzy subgroups*, Info. Sci. **73** (1993), 41 – 55.

[228] _____, *Optimal fuzzy objects for he set of given data in the case of the group theory*, Info. Sci. **92** (1996), 197 – 210.

[229] Nobuaki Kuroki, *On fuzzy ideals and fuzzy bi-ideals in semigroups*, Fuzzy Sets and Systems **5** (1981), 203 – 215.

[230] _____, *Fuzzy semiprime ideals in semigroups*, Fuzzy Sets and Systems **8** (1982), 71 – 80.

[231] _____, *On fuzzy semigroups*, Info. Sci. **53** (1991), 203 –236.

[232] _____, *Fuzzy congruences and fuzzy normal subgroups*, Info. Sci. **60** (1992), 247 – 259.

[233] _____, *Fuzzy interior ideals in semirings*, J. Fuzzy Math. **3** (1995), 435 –447.

[234] _____, *Fuzzy congruences on inverse semigroups*, Fuzzy Sets and Systems **87** (1997), 335 – 340.

[235] _____, *Rough ideals in semigroups*, Info. Sci. **100** (1997), 139 – 163.

[236] _____ & John Mordeson, *Successor and source functions*, J. Fuzzy Math. **5** (1997), 173 –182.

[237] _____ & Paul Wang, *The lower and upper approximations in a fuzzy group*, Info. Sci. **90** (1996), 203 – 220.

[238] Nobuaki Kurski, *On fuzzy ideals and fuzzy bi-ideals in semigroups*, Fuzzy Sets and Systems **5** (1981), 203 – 215.

[239] Y. Lafont & T. Streicher, *Games semantics for linear algebra*, Proc. 6th Annual IEEE Symposium on Logic in Computer Science, IEEE Computer Society Press, 1991, pp. 43 – 49.

[240] John Lake, *Sets, fuzzy sets, multisets, and functions*, J. London Math. Soc. (2) **12** (1976), 323 – 326.

[241] François Lamarche, *Quantitative domains and infinitary algebras*, Theor. Comp. Sci. **94** (1992), 37 – 62.

[242] E. Stanley Lee & Qing Zhu, *Fuzzy and Evidence Reasoning*, Physica-Verlag, Heidelberg, 1995.

[243] Stephan Lehmke, *Some properties of fuzzy ideals in a lattice*, Proceedings of the 6th IEEE International Conference on Fuzzy Systems, vol. 2, IEEE Computer Society Press, 1997, pp. 813 – 818.

[244] Yaakov Levitzki, *On powers with transfinite exponents*, Riveon LeMathematika **1** (1946), 8 – 13.

[245] Shenglin Li, Yangdong Yu & Zhudeng Wang, *T-congruence L-relations on groups and rings*, Fuzzy Sets and Systems **92** (1997), 365 – 381.

[246] Su-Yun Li, De-Gang Chen, Wen-Xiang Gu & Hui Wang, *Fuzzy homomorphisms*, Fuzzy Sets and Systems **79** (1996), 235 – 238.

[247] Grigori L. Litvinov & Victor P. Maslov, *The correspondence principle for idempotent calculus and some computer applications*, Idempotency (Bristol 1994) (J. Gunawardena, ed.), Publ. Newton Inst. #11, Cambridge Univ. Press, Cambridge, 1998, pp. 420 – 443.

[248] Liu Wang-jin, *Fuzzy invariant subgroups and fuzzy ideals*, Fuzzy Sets and Systems **8** (1982), 133 – 139.

[249] _____, *Operations on fuzzy ideals*, Fuzzy Sets and Systems **11** (1983), 31 – 41.

[250] Liu Ying-Ming, *Structures of fuzzy order homomorphisms*, Fuzzy Sets and Systems **21** (1987), 43 – 51.

[251] Daniel Loeb, *Sets with a negative number of elements*, Advances in Math **91** (1992), 64 –74.

[252] Sergio López-Permouth & D. S. Malik, *On categories of fuzzy modules*, Info. Sci. **52** (1990), 211 – 220.

[253] R. Lowen, *Fuzzy topological spaces and fuzzy compactness*, J. Math. Anal. Appl. **56** (1976), 621–633.

[254] Lu Tu & Gu Wen-Xiang, *Some properties of fuzzy vector spaces*, J. Fuzzy Math. **3** (1995), 921 – 931.

[255] P. Lubczonok, *Fuzzy vector spaces*, Fuzzy Sets and Systems **38** (1990), 329 – 345.

[256] Ian Madconald, *The Theory of Groups*, Clarendon Press, Oxford, 1968.

[257] B. Mahr, *Iteration and summability in semirings*, Ann. Discrete Math. **19** (1984), 229 – 256.

[258] Michael G. Main & David B. Benson, *Free semiring-representations and nondeterminism*, J. Comp. Sci. Sys. **30** (1985), 318 – 328.

[259] Jean Mairesse, *A graphical approach to the spectral theory in the $(max, +)$ algebra*, IEEE Transactions on Automatic Control **20** (1985), 1 – 6.

[260] B. B. Makamba, *Direct products and isomorphism of fuzzy semigroups*, Info. Sci. **65** (1992), 33 – 43.

[261] D. S. Malik & John N. Mordeson, *Fuzzy vector spaces*, Info. Sci. **55** (1991), 271 – 281.

[262] _____, *Fuzzy relations on rings and groups*, Fuzzy Sets and Systems **43** (1991), 117 – 123.

[263] _____, *Fuzzy homomorphisms of rings*, Fuzzy Sets and Systems **46** (1992), 139 – 146.

[264] D. S. Malik, John N. Mordeson & M. K. Sen, *Fuzzy normal subgroups, solvability of fuzzy subgroups and congruence relations*, J. Fuzzy Math. **2** (1994), 397 – 408.

[265] Ernest G. Manes, *Monoids, matrices, and generalized dynamic algebra*, Categorial Methods in Computer Science (H. Ehrig et al., eds.), Lecture Notes in Computer Science # 398, Springer Verlag, Berlin, 1989.

[266] _____ & D. B. Benson, *The inverse semigroup of a sum-ordered semiring*, Semigroup Forum **31** (1985), 129 – 152.

[267] Milan Mareš, *Weak arithmetics of fuzzy numbers*, Fuzzy Sets and Systems **91** (1997), 143 – 153.

REFERENCES

[268] Luis Martinez, *Fuzzy subgroups of fuzzy groups and fuzzy ideals of fuzzy rings*, J. Fuzzy Math. **3** (1995), 833 – 849.

[269] J. P. Mascle, *Quelques résultats de décidabilité sur la finitude des semigroupes de matrices*, (preprint, 1985).

[270] M. Mashinchi & M. Mukaidono, *Generalized fuzzy quotient subgroups*, Fuzzy Sets and Systems **74** (1995), 245 – 257.

[271] Victor P. Maslov, *Méthodes Opératorielles*, Mir, Moscow, 1987.

[272] _____ & S. N. Sambourskiĭ, *Idempotent Analysis*, Advances in Soviet Mathematics # 13, American Mathematical Society, Providence, 1992.

[273] G. Mayor, *Contribució a l'estudi de models matemàtics per a la lògica de la vaguetat*, PhD. thesis, Universitat de le Illes Balears, Palma de Mallorca, 1984.

[274] R. G. McLean & H. Kummer, *Fuzzy ideals in semigroups*, Fuzzy Sets and Systems **48** (1992), 137 – 140.

[275] Karl Menger, *Statistical metrics*, Proc. Nat. Acad. Sci. USA **28** (1942), 327 – 335.

[276] _____, *Probabilistic theories of relations*, Proc. Nat. Acad. Sci. USA **37**, 178 – 180.

[277] Radko Mesiar, *Toll connectives*, J. Fuzzy Math. **1** (1993), 327 – 335.

[278] _____, *A note on the T-sum of L-R fuzzy numbers*, Fuzzy Sets and Systems **79** (1996), 259 – 261.

[279] _____, *Triangular-norm-based addition of fuzzy intervals*, Fuzzy Sets and Systems **91** (1997), 231 – 237.

[280] _____ & Endre Pap, *Different interpretations of triangular norsm and related operations*, Fuzzy Sets and Systems **96** (1998), 183 – 189.

[281] _____ & Ján Rybárik, *Pan-operations structure*, Fuzzy Sets and Systems **74** (1995), 365 – 369.

[282] Sidney S. Mitchell & Porntip Sinutoke, *The theory of semifields*, Kyungpook Math. J. **22** (1982), 325 – 348.

[283] S. Miyamoto, *Fuzzy Sets in Information Retrieval and Cluster Analysis*, Kluwer, Dordrecht, 1990.

[284] _____, *Complements, t-norms, and s-norms in toll sets*, Proceedings of the 2nd International Conference on Fuzzy Logic and Neural Networks, (Iizuka, Japan, July 17-22, 1992), 1992, pp. 579 – 582.

[285] Masaharu Mizumoto & K. Tanaka, *Some properties of fuzzy sets of type* 2, Inform. and Control **31** (1976), 312 – 340.

[286] _____, *Fuzzy sets and their operations*, Inform. and Control **48** (1981), 30 – 48.

[287] Ramon E. Moore, *Interval Analysis*, Prentice-Hall, Englewood Cliffs, N. J., 1966.

[288] John N. Mordeson, *Generating properties of fuzzy algebraic structures*, Fuzzy sets and systems **55** (1993), 107 – 120.

[289] _____, *Bases of fuzzy vector spaces*, Info Sci. **67** (1993), 87 – 92.

[290] Nehad N. Morsi & Samy El-Badawy Yehia, *Fuzzy-quotient groups*, Info. Sci. **81** (1994), 177 – 191.

[291] Godfrey C. Muganda, *Free fuzzy modules and their bases*, Info. Sci. **72** (1993), 65 – 82.

[292] A. Muir & M. W. Warner, *Lattice valued relations and automata*, Discrete Appl. Math **7** (1984), 65 – 78.

[293] T. K. Mukherjee & P. Bhattacharya, *Fuzzy groups: some group theoretic analogs*, Info. Sci. **39** (1986), 247 – 268.

[294] T. K. Mukherjee & M. K. Sen, *On fuzzy ideals of a ring I*, Fuzzy Sets and Systems **21** (1987), 99 – 104.

[295] V. Murali, *Fuzzy equivalence relations*, Fuzzy sets and systems **30** (1989), 155 - 163.

[296] N. V. E. S. Murthy, *Hypergraphs and fuzzy languages*, (preprint, 1993).

[297] _____, *How fuzzy are the fuzzy ideals,*, J. Fuzzy Math. **4** (1996), 15 – 23.

[298] C. V. Negoiță & Dan A. Ralescu, *Applications of Fuzzy Sets to Systems Analysis*, Birkhäuser, Basel, 1975.

[299] William C. Nemitz, *Fuzzy relations and fuzzy functions*, Fuzzy Sets and Systems **19** (1986), 177 – 191.

[300] Hung T. Nguyen & Elbert A. Walker, *A First Course in Fuzzy Logic*, CRC Press, Boca Raton, 1997.

[301] Jan Okniński, *Semigroup Algebras*, Marcel Dekker, New York, 1990.

[302] Geert Jan Olsder, *Eigenvalues of dynamic max-min systems*, Discrete Event Dynamical Systems: Theory and Applications **1** (1991), 177 – 207.

[303] _____, *About difference equations, algebras and discrete events*, Topics in Engineering Mathematics (A. van der Burgh & J. Simonis, eds.), Kluwer, Dordrecht, 1992, pp. 121 – 150.

[304] S. Orlovski, *Calculus of Decomposable Properties, Fuzzy Sets, and Decisions*, Allerton Press, New York, 1994.

[305] Sergei V. Ovchinnikov, *Structure of fuzzy binary relations*, Fuzzy Sets and Systems **6** (1981), 169 – 175.

[306] _____, *Similarity relations, fuzzy partitions and fuzzy orderings*, Fuzzy Sets and Systems **40** (1991), 107 – 126.

[307] _____, *The duality principle in fuzzy set theory*, Fuzzy Sets and Systems **42** (1991), 133 – 144.

[308] _____, *On the image of a fuzzy subgroup*, Fuzzy Sets and Systems **81** (1996), 235 – 236.

[309] _____, *On the image of an L-fuzzy group*, Fuzzy Sets and Systems **94** (1998), 129 – 131.

[310] Fuzheng Pan, *The various structures of fuzzy quotient modules*, Fuzzy Sets and Systems **50** (1992), 187 – 192.

[311] Zoltan Papp, *On the strong injective (projective) dimension of modules*, Arch. Math. **25** (1974), 354 – 360.

[312] Rohit Parikh, *Some applications of topology to program semantics*, Logic of Programs (D. Kozen, ed.), Lecture Notes in Computer Science #131, Springer-Verlag, Berlin, 1982, pp. 375 – 386.

[313] J. Pavelka, *On fuzzy logic I (Many-valued rules of inference)*, Zwitschr. math. Logic und Grundagen Math. **25** (1979), 45 – 52.

[314] S. Pedrycz, *Fuzzy Control and Fuzzy Systems*, Report # 82, Department of Mathematics, Delft University of Technology, 1982.

[315] Witold Pedrycz, *On generalized fuzzy relational equations and their applications*, J. Math. and Appl. **107** (1985), 520 – 536.

[316] _____, *Processing in relational structures: fuzzy relational equations*, Fuzzy Sets and Systems **40** (1991), 77 – 106.

[317] _____, *Fuzzy modelling fundamentals, construction and evaluation*, Fuzzy Sets and Systems **41** (1991), 1 – 15.

[318] _____, *s-t Fuzzy relational equations*, Fuzzy Sets and Systems **59** (1993), 189 – 195.

[319] Keti Peeva, *On algebraic structures for matrices, relations and graphs over a semiring*, Mat. Bull. **7-8** (1983-4), 36-48.

[320] Jean-Eric Pin, *Tropical semirings*, Idempotency (Bristol 1994) (J. Gunawardena, ed.), Publ. Newton Inst. #11, Cambridge Univ. Press, Cambridge, 1998, pp. 50 – 69.

[321] A. M. Pitts, *Fuzzy sets do not form a topos*, Fuzzy Sets and Systems **8** (1982), 101 – 104.

[322] J. S. Ponizovskiĭ, *Semigroup rings*, Semigroup Forum **36** (1987), 1 – 46.

[323] Vaughan Pratt, *Modeling concurrency with partial orders*, Int. J. Parallel Processing **15** (1986), 33 – 71.

[324] _____, *The second calculus of binary relations*, MFCS '93, Lecture Notes in Computer Science #711, Springer-Verlag, Berlin, 1993, pp. 142 – 155.

[325] _____, *Chu spaces, complementarity and uncertainty in rational mechanics*, (preprint, 1994).

[326] _____, *The Stone gamut: a coordinatization of mathematics*, Proc. 10th Annual Symposium on Logic in Computer Science, IEEE Computer Society Press, 1995, pp. 444 – 454.

[327] _____, *Chu spaces and their interpretation as concurrent objects*, Computer Science Today, Lecture Notes in Computer Science #1000, Springer-Verlag, Berlin, 1995, pp. 392 – 405.

[328] _____, *Reconciling event structures and higher dimensional automata*, (preprint, 1996).

[329] _____, *Chu spaces from the representational viewpoint*, (preprint, 1997).

[330] Helena Rasiowa & Nguyen Cat Ho, *LT-fuzzy sets*, Fuzzy Sets and Systems **47** (1992), 323 – 339.

[331] Suryansu Ray, *Isomorphic fuzzy groups*, Fuzzy Sets and Systems **50** (1992), 201 – 207.

[332] _____, *Classification of fuzzy subgroupoids of a group (I)*, Info. Sci. **73** (1993), 57 – 76.

[333] _____, *Classification of fuzzy subgroupoids of a group (II)*, Info. Sci. **73** (1993), 77 – 91.

[334] _____, *Interaction between the fuzzy subsets and the endomorphisms of a group*, Info. Sci. **75** (1993), 35 – 45.

[335] _____, *Solvable fuzzy groups*, Info. Sci. **75** (1993), 47 – 61.

[336] _____, *The lattice of all idempotent fuzzy subsets of a groupoid*, Fuzzy Sets and Systems **96** (1998), 239 – 245.

[337] Rimhak Ree, *Lie elements and an algebra associated with shuffles*, Ann. Math. **68** (1958), 210–220.

[338] Nicholas Rescher, *Many-Valued Logic*, McGraw Hill, New York, 1969.

[339] Christophe Reutenauer, *Free Lie Algebras*, Clarendon Press, Oxford, 1993.

[340] Beloslav Riečan & Tibor Neubrunn, *Integral, Measure and Ordering*, Kluwer, Dordrecht, 1997.

[341] Christian Ronse & Henk J. A. M. Heijmans, *A lattice-theoretical framework for annular filters in morphological image processing*, Applicable Algebra in Engineering, Communications and Computing **9** (1998), 45 – 89.

[342] Azriel Rosenfeld, *Fuzzy groups*, J. Math. Anal. Appl. **35** (1971), 512 – 517.

[343] _____, *Fuzzy graphs*, Fuzzy Sets and Their Applications to Cognitive and Decision Process (Lotfi Zadeh et al., eds.), Academic Press, New York, 1975, pp. 77 – 95.

[344] Ivo Rosenberg, *Algebraic properties of a general convolution*, Algebraic, Extremal and Metric Combinatorics, 1986 (M.-M. Deza et al., eds.), London Math. Soc. Lecture Note Series #131, Cambridge University Press, Cambridge, 1988.

[345] Gian-Carlo Rota, *Finite Operator Calculus*, Academic Press, New York, 1978.

[346] Arto Salomaa & Matti Soittola, *Automata-theoretic Aspects of Formal Power Series*, Springer-Verlag, Berlin, 1978.

[347] S. N. Samborskiĭ & A. A. Tarashchan, *Semirings occuring in multicriteria optimization problems and in analysis of computing media*, Soviet Math. Dokl. 40 (1990), 441 – 445.

[348] _____, *The Fourier transform and semirings of Pareto sets*, Idempotent Analysis, Advances in Soviet Mathematics # 13, American Mathematical Society, Providence, 1992, pp. 139 – 150.

[349] Marouf A. Samhan, *Fuzzy congruences on semigroups*, Info. Sci. 74 (1993), 165 – 175.

[350] _____, *Fuzzy quotient algebras and fuzzy factor congruences*, Fuzzy Sets and Systems 73 (1995), 269 – 277.

[351] Elie Sanchez, *Resolution of composite fuzzy relation equations*, Inform. and Control 30 (1976), 38 – 48.

[352] _____, *Fuzzy relation equations: methodology and applications*, Fuzzy Sets, Theory and Applications (André Jones, Arnold Kaufmann & Hans-Jürgen Zimmermann, eds.), Reidel, Dordrecht, 1985.

[353] Moritoshi Sasaki, *Fuzzy functions*, Fuzzy Sets and Systems 55 (1993), 295 – 301.

[354] P. K. Saxena, *Fuzzy subgroups as union of two fuzzy subgroups*, Fuzzy Sets and Systems 57 (1993), 209 – 218.

[355] Norbert Schmechel, *On lattice-isomorphism between equivalence relations and fuzzy partitions*, Proceedings of the 25th International Symposium on Multiple-Valued Logic, IEEE Computer Society Press, 1995, pp. 146 – 151.

[356] B. Schweizer & A. Sklar, *Statistical metric spaces*, Pacific J. Math 10 (1960), 313 – 334.

[357] _____, *Probabilistic Metric Spaces*, North-Holland, Amsterdam, 1983.

[358] Dana Scott, *Data types as lattices*, SIAM J. Comput. 5 (1976), 522 – 587.

[359] J. Serra, *Image Analysis and Mathematical Morphology*, Academic Press, London, 1982.

[360] G. Shafer, *A Mathematical Theory of Evidence*, Princeton University Press, Princeton, 1976.

[361] Howard Sherwood, *Products of fuzzy subgroups*, Fuzzy Sets and Systems 11 (1983), 79 – 89.

[362] M. A. Shubin, *Algebraic remarks on idempotent semirings and the kernel theorem in spaces of bounded functions*, Idempotent Analysis, Advances in Soviet Mathematics #13, American Mathematical Society, Providence, 1992, pp. 151 – 166.

[363] Harold Simmons, *The semiring of topologizing filters of a ring*, Israel J. Math. 61 (1988), 217 – 284.

[364] Imre Simon, *Limited subsets of the free monoid*, Proc. of the 19th Annual Symposium on Foundations of Computer Science, IEEE Computer Society Press, 1978, pp. 143 – 150.

[365] _____, *Recognizable sets with multiplicities in the tropical semiring*, Mathematical Foundations of Computer Science 1988 (M. P. Chytil et al., eds.), Lecture Notes in Computer Science # 324, Springer-Verlag, Berlin, 1988, pp. 107 – 120.

[366] _____, *On semigroups of matrices over the tropical semiring*, RAIRO Inform. Theor. Appl. 28 (1994), 277 – 294.

[367] Robert de Simone, *Languages infinitaires et produit de mixage*, Theor. Comp. Sci. **31** (1984), 83–100.

[368] A. Skowron & J. Grzymaia-Busse, *From the rough set theory to evidence theory*, (preprint, 1991).

[369] Richard Squire, *The fundamental completeness of a K-finite object of truth values in a topos*, (preprint, 1994).

[370] M. V. Subbarao, *On some arithmetical convolutions*, The Theory of Arthmetical Functions (Anthony A. Gioia & Donald L. Goldsmith, eds.), Lecture Notes in Mathematics #251, Springer-Verlag, Belin, 1972, pp. 247 – 271.

[371] Philipp Sünderhauf, *Tensor products and powerspaces in quantitative domain theory*, Electronic Notes in Theoretical Computer Science **6** (1997).

[372] Yasushi Suzuki, *On the construction of free fuzzy groups*, J. Fuzzy Math. **2** (1994), 1 – 15.

[373] U. M. Swamy & N. V. E. S. Murty, *Representation of fuzzy subalgebras by crisp subalgebras*, (preprint, 1995).

[374] U. M. Swamy & D. Viswanadha Raju, *Algebraic fuzzy systems*, Fuzzy Sets and Systems **41** (1991), 187 – 194.

[375] T. Tamura & J. Shafer, *Power semigroups*, Math. Japon. **12** (1967), 25 –32.

[376] Toshiro Terano, Kiyoji Asai & Michio Sugeno, *Fuzzy Systems Theory and Its Applications*, Academic Press, London, 1987.

[377] Yutaka Terao & Naoki Kitsunezaki, *Fuzzy sets and linear mappings on vector spaces*, Math. Japon. **39** (1994), 61 – 68.

[378] Bret Tilson, *Monoid kernels, semidirect products, and their adjoint relation*, Monoids and Semigroups with Applications (John Rhodes, ed.), World Scientific, Singapore, 1991, pp. 31 – 54.

[379] H. Toth, *Categorical properties of f-set theory*, Fuzzy Sets and Systems **33** (1989), 99 – 109.

[380] Enric Trillas, C. Alsina & L. Valverde, *Do we need max, min, and $1-j$ in fuzzy set theory?*, Fuzzy Set and Possibility Theory (R. R. Yager, ed.), Pergamon Press, New York, 1982, pp. 275 – 297.

[381] Enric Trillas, Susana Cubillo & Cristina del Campo, *A few remarks on some T-conditional functions*, Proceedings of the 6th IEEE International Conference on Fuzzy Systems, vol. 1, IEEE Computer Society Press, 1997, pp. 153 – 156.

[382] A. S. Troelstra, *Lectures on Linear Logic*, CSLI Lecture Notes No. 29, Center for the Study of Language and Information, Stanford, Calif., 1992.

[383] Esko Turunen, *Algebraic structures in fuzzy logic*, Fuzzy Sets and Systems **52** (1992), 181 – 188.

[384] _____, *Well-defined fuzzy sentential logic*, Math. Log. Quart. **41** (1995), 236 – 248.

[385] _____, *BL-algebras of basic fuzzy logic*, (preprint, 1997).

[386] Dragos Vaida & Alexandru Matescu, *Towards a unified theory of sequential, parallel and semi-parallel processes*, (preprint, 1993).

[387] L. Valverde, *On the structure of F-indistinguishability operators*, Fuzzy Sets and Systems **17** (1985), 313 – 328.

[388] S. Vickers, *Topology via Logic*, Cambridge University Press, Cambridge, 1989.

[389] K. A. Volosov & V. P. Maslov, *Mathematical Aspects of Computer Engineering*, Mir, Moscow, 1988.

[390] J. Vrba, *General decomposition problem of fuzzy relations*, Fuzzy Sets and Systems **54** (1993), 69 – 79.

[391] Vu Ha & Peter Haddawy, *Geometric foundations for interval-based probabilities*, (preprint, 1998).

[392] Michael Wagenknecht, *On some relations between fuzzy similarities and metrics under Archimedian t-norms*, J. Fuzzy Math. **3** (1995), 563 – 572.

[393] Abraham Wald, *On a statistical generalization of metric spaces*, Proc. Nat. Acad. Sci. USA **29** (1943), 196 – 197.

[394] Wang Huaxiong, *On characters of semirings*, (preprint, 1995).

[395] Wang Jin-Liu, *Fuzzy invariant subgroups and fuzzy ideals*, Fuzzy Sets and Systems **8** (1982), 133 – 139.

[396] Wang Zhenyuan, *Fuzzy convolution of fuzzy distributions on groups*, Analysis of Fuzzy Information, vol. I (James C. Bezdek, ed.), CRC Press, Boca Raton, 1986, pp. 97 – 103.

[397] Wang Zhudeng, *TL-submodules and TL-linear subspaces*, Fuzzy Sets and Systems **68** (1994), 211 – 225.

[398] M. W. Warner & A. Muir, *Everything is fuzzy*, The Mathematics of Fuzzy Systems (Antonio Di Nola & Aldo Ventre, eds.), Verlag TÜV Rheinland, Köln, 1986, pp. 303 – 323.

[399] Siegfried Weber, *A general concept of fuzzy connectives, negations and implications based on t-norms and t-conorms*, Fuzzy Sets and Systems **11** (1983), 115 – 134.

[400] W. Wechler, *Hoare algebras versus dynamic algebras*, Algebra, Combinatorics and Logic (J. Demetrovics et al., eds.), Colloquia Math. Soc. János Bolyai #42, North Holland, Amsterdam, 1986.

[401] _____, *R-fuzzy computation*, J. Math. Anal. Appl. **115** (1986), 225–232.

[402] Friedrich Wehrung, *Injective positively ordered monoids I*, J. Pure Appl. Alg. **83** (1992), 43 – 82.

[403] K. Weihrauch & Ulrich Schreiber, *Embedding metric spaces in CPO's*, Theor. Comp. Sci. **16** (1981), 5 – 24.

[404] Richard Wiegandt, *On the general theory of Möbius inversion formula and Möbius product*, Acta Sci. Math. (Szeged) **20** (1959), 164 – 180.

[405] Jósef Winkowski, *Concatenable weighted pomsets and their applications to modelling processes of Petri nets*, Fund. Inform. **28** (1996), 403 – 421.

[406] Z. Wong & G. J. Klir, *Fuzzy Measure Theory*, Plenum, New York, 1992.

[407] Ahnont Wongseelashote, *Semirings and path spaces*, Discrete Math. **26** (1979), 55 – 78.

[408] Wu Fuming, *Fuzzy time semirings and fuzzy-timing colored Petri nets*, (preprint, 1998).

[409] _____, *A Framework for Dynamical Modelling of Information Systems: the R-net approach*, PhD thesis, University of Haifa (1998).

[410] Ying Xie, *ω-Complete Semirings and Matrix Iteration Theories*, PhD thesis, Stevens Institute of Technology (1991).

[411] Ronald R. Yager, *Aggregation operators and fuzzy systems modeling*, Fuzzy Sets and Systems **67** (1994), 129 – 145.

[412] Y. Y. Yao & S. K. M. Wong, *Interval approaches for uncertain reasoning*, Foundations of Intelligent Systems, Proceedings of ISMIS'97 (Z. W. Raś & A. Skowron, eds.), Lecture Notes in Computer Science #1325, Springer-Verlag, Berlin, 1997, pp. 381 – 390.

[413] R. T. Yeh, *Toward an algebraic theory of fuzzy rational relations*, Proceedings of the International Congress of Cybernetics, Namur, Belgium, 1973, pp. 205 – 223.

[414] Yu Yandong, *Finitely generated T-fuzzy linear spaces*, Fuzzy Sets and systems **30** (1989), 69 – 81.

[415] _____ & Zhudeng Wang, *TL-subrings and TL-ideals. Part 1. Basic concepts*, Fuzzy Sets and Systems **68** (1994), 93 – 103.

[416] Lofti Zadeh, *Fuzzy Sets*, Inform. and Control **8** (1965), 338–353.

[417] _____, *Similarity relations and fuzzy orderings*, Info. Sci. **3** (1971), 177 – 200.

[418] _____, *Outline of a new approach to the analysis of complex systems and decision processes*, IEEE Trans. Systems, Man. and Cybernetics **3** (1973), 28 – 44.

[419] _____, *The concept of linguistic variable and its applications to approximate reasoning. I*, Info. Sci. **8** (1975), 199 - 251.

[420] _____, *The concept of linguistic variable and its applications to approximate reasoning. II*, Info. **8** (1975), 301 – 357.

[421] _____, *The concept of linguistic variable and its applications to approximate reasoning. III*, Info. Sci. **9** (1975), 43 – 80.

[422] M. M. Zahedi, *Some results on L-fuzzy modules*, Fuzzy Sets and Systems **53** (1993), 355 – 361.

[423] _____, *Upper-continuous and compactly generated properties of lattice of L-fuzzy modules*, Fuzzy Sets and Systems **60** (1993), 103 – 108.

[424] Hans Zassenhaus, *The Theory of Groups*, Chelsea, New York, 1958.

[425] Yunjie Zhang & Kaiqi Zou, *A note on an equivalence relation on fuzzy subgroups*, Fuzzy Sets and Systems **95** (1998), 243 – 247.

[426] Zhao Jianli, *The categories of fuzzy modules*, J. Fuzzy Math. **4** (1966), 491 – 501.

[427] _____, Shi Kaiquan & Yue Mingshan, *Fuzzy modules over fuzzy rings*, J. Fuzzy Math **1** (1993), 531 –539.

[428] Uwe Zimmermann, *Linear and Combinatorial Optimization in Ordered Algebraic Structures*, North Holland, Amsterdam, 1981.

Index

A

action 28
additive inverse 25
additive set of generators 64
additively-idempotent semiring 11
adjoint
—, left 42
—, right 42
admissible family of functions 14
aggregation 12
algebra
—, command 11
—, incline 8
—, information 12
— $(max, +)$- 17
—, multiple-valued 19
— of formal power series 70
—, power 67
—, restricted convolution 68
—, schedule 17
—, semiring-valued incidence 41
—, syntactic 73
—, Wiegand convolution 43
algebraic semiring 64
algebraically-closed semiring 23
alphabet 70
amplification of signals 25
arity 87

arrow 71
assignment, basic 4
augmentation
— congruence 81
— ideal 74
— morphism 81
auxiliary symbols 87

𝔹
bag 2
Bandler-Kohout composition 40
basic
— assignments 4
— semiring 7
basis 149
— for a function 57
belt 11
bounded function 33

ℂ
Cauchy product 70
central function 76
character 61
characteristic
— function 1
— function with values in a semiring 33
— of a semiring 8
Chu space 38
clan 31
classical semiring of fractions 24
CLO-semiring 16
closed function 56
closure operator 56
—, Fréchet 49
—, linear 56
—, transitive 82
command algebra 11
commutative monoid, free 29
compact element of a semiring 20

compatible
— quasimetrics 54
— relations 94
— semiring-valued relations 51
complementation 13
complete
— class of subsets 4
— gradation 57
— semiring 14
complete-lattice-ordered semiring 16
complex 69
composition
—, Bandler-Kohout 40
— of signals, parallel 25
congruence, augmentation 81
— class 110
— relation on a semiring 24
— semiring-valued relation 108
—, syntactic 73
conjugate 114
conorm, triangular 3, 4, 12
constraint 29
continuous time scale ....
convex geometry 50
convolution 67
— algebra, restricted 68
— algebra, Wiedgandt 43
— context 68
cost of membership function 3
counter time scale 16
crisp
— function 63
— subset 28

𝔻
data structure, underlying 28
decreasing subset 15
difference-ordered semiring 9
dioïd 10

directed subset 98
distance distribution function 72
divisibility monoid 12
division semiring 20
domain 57
—, semantic 69

$\mathbb{E}$

entire semiring 12
equality relation, with respect to a semiring 51
equivalence
— class, with respect to a semiring 51
—, Lipschitz 48
— relation, semiring-valued 46
— relation, semiring-valued with weight 52
— relation, strong semiring-valued 46
evaluation function 62, 63
— morphism 28
event 28, 38
exponent set 27
expression 87
extended
— metric 47
— pseudometric 47
— seminorm 115
extent of membership function 2

$\mathbb{F}$

f-set 3
facile
— semiring-valued subgroup 113
— semiring-valued subsemigroup 89
finitary semigroup 67
formal power series 70
fraction 22
frame-ordered semiring 18
Fréchet closure operator 49
free
— commutative monoid 29

— group ········ 116
— semiring-valued submodule ········ 146
Fubini's Theorem ········ 14
function
—, bounded ········ 33
—, central ········ 76
—, characteristic ········ 1
—, cost of membership ········ 3
—, closed ········ 57
—, crisp ········ 63
—, distance distribution ········ 72
—, evaluation ········ 62, 63
—, improper ········ 61
—, maximally free ········ 57
—, membership indicator ········ 28
—, normal ········ 32
—, point-complete ········ 34
—, predimension ········ 146
—, projection ········ 63
—, proper ········ 61
—, quasidimension ········ 145
—, semiring-valued ········ 39, 42
—, stable ········ 56
—, subadditive ········ 90
—, subnormal ········ 32
—, unimodal ········ 32
fusion product ········ 71
fuzz ········ 13
fuzzy
— game ········ 94
— number ········ 69
— set of small elements ········ 90
— subset ········ 2
— topology ········ 18

G

game ········ 38
—, fuzzy ········ 94
—, superadditive ········ 94

Gel'fand semiring ········ 8
geometry, convex ········ 50
good set of points ········ 34
gradation ········ 57
—, complete ········ 57
graded set ········ 57
group, free ········ 116

ℍ
Head's metatheorem ········ 87
height ········ 32
hemiring ········ 7
homomorphism with values in a semiring ········ 73
hybrid set ········ 2

𝕀
ideal
—, augmentation ········ 74
— of a semigroup ········ 104
—, regular semiring-valued ········ 155
—, semiring-valued ········ 155
idempotent
— analysis ········ 29
— measure ········ 21
implication ········ 44
improper function ········ 61
incidence algebra, semiring valued ········ 41
incline algebra ········ 8
information algebra ········ 12
input ········ 38
interpretation ········ 69
interval analysis ········ 3, 11
— in a semiring ········ 10
intrinsic product ········ 163
inversive semigroup ········ 104
irreducible semiring-valued subgroup ········ 120
isomorphism between semiring-valued subsemigroups ········ 107

J

Jacobi matrix ········ 73

K

kernel, weighing ········ 75
Kleene star ········ 14
Krull-Kaplansky-Jaffard-Ohm Theorem ········ 20

L

language ········ 29, 48, 70
lattice-ordered semiring ········ 11
left
— absorbing subset ········ 78
— adjoint ········ 42
— compatible relation ········ 108
— ideal of a semigroup ········ 104
— ideal, semiring-valued ········ 155
— residual ········ 19
linear closure operator ········ 56
linearly independent set ········ 149
Lipschitz equivalence ········ 48
location ········ 38
lower set ········ 10
Lukasiewicz-type residual ········ 19

M

map, semiring-valued ········ 41
$(max, +)$-algebra ········ 17
maximally free function ········ 57
Mascle semiring ········ 17
measure
—, idempotent ········ 21
—, possibility ········ 4
—, Sugeno ········ 4
membership
— cost function ········ 3
— indicator function ········ 28
metric
—, extended ········ 47
— semigroup ········ 110

— space, statistical 93
Minkowski addition 71
model 48
module 25
—, normed 142
modus ponens rule with values in a semiring 44
monoid
—, divisibility 12
—, free commutative 29
—, partial 70
— semiring 70
—, syntactic 73
morphism
—, augmentation 81
— between functions 32
— between semiring-valued subsemigroups 107
—, evaluation 28
— of complete semirings 61
— of semirings 61
multiple-valued algebra 19
multiplication, scalar 25
multiplicatively-cancellative semiring 22
multiplicity function 2
multiset 2
multisubset
—, partially-ordered 29
—, totally ordered 29
multivalued function 4

N

necessary summation 15
negation 13
node 71
nontrivial point 33
norm, triangular 3
normal
— function 32
— semiring-valued subgroup 128, 132
normalization condition 72

INDEX

normed module 142
number, fuzzy 69

𝕆

operation 87
—, power 4
opposite relation 38
order of determinacy 11
Øre set 24

ℙ

parallel composition of signals 25
partial
— monoid 70
— semigroup 70
partially-ordered multisubset 29
partition, semiring valued 46
path 71
penalty theory 4
Φ-operator 19
plinth 32
point 38
— homomorphism 107
—, nontrivial 33
—, trivial 33
— with values in a semiring 33
point-complete function 34
polarity 13
pomsubset 29
possibility measure 4
power
— algebra 67
— domain construction ....
— operation 4
— representation 69
— series, formal 70
predimension function 146
prime 65
process 29

product
—, Cauchy ········ 70
—, fusion ········ 71
—, intrinsic ········ 163
—, pseudodirect ········ 20
projection function ········ 63
proper function ········ 61
pseudodirect product ········ 20
pseudomeasure ········ 4
pseudometric
—, extended ········ 47
—, with values in a semiring ········ 53

ℚ

QLO-sdemiring ········ 17
quantic-lattice-ordered semiring ········ 17
quasidimension function ········ 145
quasi-inverse ········ 14
quasimetric, with values in a semiring ········ 52
quiver ········ 71
quotient semiring ········ 24

ℝ

range of values ········ 28
rational identity ········ 15
reducible semiring-valued subgroup ········ 120
reflextive semiring-valued relation ········ 45
regular
— semigroup ········ 105
— semiring in the sense of Von Neumann ········ 159
— semiring-valued ideal ········ 155
— semiring-valued subgroup ········ 113
— semiring-valued submodule ········ 141
— semiring-valued submonoid ········ 89
relation
—, congruence on a semiring ········ 24
—, equality with respect to a semiring ········ 51
—, equivalence semiring-valued ········ 46
—, left compatible ········ 108

—, opposite 38
—, reflexive semiring-valued 45
—, right compatible 108
—, semiring-valued 37, 38
—, strong semiring-valued equivalence 44
—, strongly-transitive semiring-valued 44
—, symmetric semiring-valued 45
—, transitive semiring-valued 44
—, weakly-symmetric semiring-valued 45
—, weakly-transitive semiring-valued 45
relations, compatible 51
representation
— of a semiring 32
—, power 69
residual
—, left 19
—, Lukasiewicz-type 19
—, right 19
resource 29, 30
right
— adjoint 42
— compatible relation 108
— ideal of a semigroup 104
— ideal, semiring-valued 155
— residual 19
rule 100

S

satisfaction 48
scalar multiplication 25
schedule algebra 17
self-normal 133
semantic domain 69
semifield 20
— of fractions 22
semigroup
—, finitary 67
—, inversive 104
—, metric 110

—. partial ········ 70
—, regular ········ 105
—, strongly metric ········ 110
semimodule ········ 25
—, zerosumfree ········ 25
seminorm, extended ········ 115
semiring ········ 5, 7
—, additively-idempotent ········ 11
—, algebraic ········ 64
—, algebraically closed ········ 23
—, basic ········ 7
—, CLO- ········ 16
—, complete ········ 14
—, complete-lattice-ordered ········ 16
—, difference-ordered ········ 9
—, division ········ 20
—, entire ········ 12
—, frame-ordered ········ 18
—, Gel'fand ········ 8
—, lattice-ordered ········ 11
—, Mascle ········ 17
—, monoid ········ 70
—, multiplicatively cancellative ········ 22
— of fractions, classical ········ 24
—, QLO- ········ 17
—, quantic-lattice-ordered ········ 17
—, quotient ········ 24
—, regular in the sense of Von Neumann ········ 159
—, simple ········ 8
—, tropical ········ 16
—, zerosumfree ········ 10
semi-Thue system ········ 100
set of generators, additive ········ 64
signal ········ 25
simple semiring ········ 8
simply-generated semiring-valued subspace ········ 153
simulation ········ 42
slope ········ 11
small elements ········ 90

stable function ········ 56
state ········ 37, 38, 48
statistical metric space ········ 93
string ········
strong semiring-valued equivalence relation ········ 46
strongly
— metric semigroup ········ 110
— transitive semiring-valued relation ········ 44
subadditive
— function ········ 90
— semiring-valued subsemigroup ········ 90
subgroup
—, semiring-valued ········ 113
—, irreducible ········ 120
—, normal ········ 128
—, reducible ········ 120
submodule
—, semiring-valued ········ 141
— semiring-valued free ········ 146
— semiring-valued generated by a subset ········ 145
— semiring-valued generated by finitely-many points ········ 145
—, regular semiring-valued ········ 141
submonoid
—, facile semiring-valued ········ 89
—, regular semiring-valued ········ 89
—, semiring-valued ········ 89
subnormal function ········ 32
subsemigroup
—, facile semiring-valued ········ 113
—, generated by a function ········ 102
—, regular semiring-valued ········ 113
—, self-normal semiring-valued ········ 133
—, semiring-valued ········ 89
—, subadditive ········ 90
subsemimodule ········ 25
—, with values in a semiring ········ 95
subset
—, crisp ········ 28
—, decreasing ········ 15

—, directed 98
—, fuzzy 2
—, left absorbing 78
—, subtractive 10
—, toll 3
subspace
—, semiring-valued 147
—, simply-generated semiring-valued 153
subtractive subset 10
Sugeno measure 4
summation 14
—, necessary 15
superadditive fuzzy game 94
supervisors of automata 4
support 27
symbol, auxiliary 87
symmetric semiring-valued relation 45
syntactic
— algebra 73
— congruence 73
— monoid 73

T
task 29
TETRIS 30
theory 48
time scale
—, continuous 17
—, counter 16
toll
— subset 3
— topology 18
tomsubset 29
topology
—, fuzzy 18
—, toll 18
totally-ordered multisubset 29
trace 42
transform 42

transitive
— closure with respect to a convolution ········ 82
— semiring-valued relation ········ 44
triangular
— conorm ········ 3, 4, 12
— norm ········ 3
tribe ········ 31
trivial point ········ 33
tropical semiring ········ 16

𝕌
underlying data structure ········ 28
unimodal function ········ 32

𝕍
value ········ 38
variable ········ 38, 87

𝕎
weakly-symmetric semiring-valued relation ········ 45
weakly-transitive semiring-valued relation ········ 45
weight ········ 135
— of a semiring-valued equivalence relation ········ 52
weighing kernel ········ 75
well-ordered function ········
Wiedgandt convolution algebra ········ 43

ℤ
zeroid ········ 9
zerosumfree
— semimodule ········ 25
— semiring ········ 10

Other *Mathematics and Its Applications* titles of interest:

Y. Fong, H.E. Bell, W.-F. Ke, G. Mason and G. Pilz (eds.): *Near-Rings and Near-Fields*. 1995, 278 pp. ISBN 0-7923-3635-6

A. Facchini and C. Menini (eds.): *Abelian Groups and Modules*. (Proceedings of the Padova Conference, Padova, Italy, June 23–July 1, 1994). 1995, 537 pp. ISBN 0-7923-3756-5

D. Dikranjan and W. Tholen: *Categorical Structure of Closure Operators*. With Applications to Topology, Algebra and Discrete Mathematics. 1995, 376 pp.
ISBN 0-7923-3772-7

A.D. Korshunov (ed.): *Discrete Analysis and Operations Research*. 1996, 351 pp.
ISBN 0-7923-3866-9

P. Feinsilver and R. Schott: *Algebraic Structures and Operator Calculus*. Vol. III: Representations of Lie Groups. 1996, 238 pp. ISBN 0-7923-3834-0

M. Gasca and C.A. Micchelli (eds.): *Total Positivity and Its Applications*. 1996, 528 pp.
ISBN 0-7923-3924-X

W.D. Wallis (ed.): *Computational and Constructive Design Theory*. 1996, 368 pp.
ISBN 0-7923-4015-9

F. Cacace and G. Lamperti: *Advanced Relational Programming*. 1996, 410 pp.
ISBN 0-7923-4081-7

N.M. Martin and S. Pollard: *Closure Spaces and Logic*. 1996, 248 pp.
ISBN 0-7923-4110-4

A.D. Korshunov (ed.): *Operations Research and Discrete Analysis*. 1997, 340 pp.
ISBN 0-7923-4334-4

W.D. Wallis: *One-Factorizations*. 1997, 256 pp. ISBN 0-7923-4323-9

G. Weaver: *Henkin–Keisler Models*. 1997, 266 pp. ISBN 0-7923-4366-2

V.N. Kolokoltsov and V.P. Maslov: *Idempotent Analysis and Its Applications*. 1997, 318 pp.
ISBN 0-7923-4509-6

J.P. Ward: *Quaternions and Cayley Numbers*. Algebra and Applications. 1997, 250 pp.
ISBN 0-7923-4513-4

E.S. Ljapin and A.E. Evseev: *The Theory of Partial Algebraic Operations*. 1997, 245 pp.
ISBN 0-7923-4609-2

S. Ayupov, A. Rakhimov and S. Usmanov: *Jordan, Real and Lie Structures in Operator Algebras*. 1997, 235 pp. ISBN 0-7923-4684-X

A. Khrennikov: *Non-Archimedean Analysis: Quantum Paradoxes, Dynamical Systems and Biological Models*. 1997, 389 pp. ISBN 0-7923-4800-1

G. Saad and M.J. Thomsen (eds.): *Nearrings, Nearfields and K-Loops*. (Proceedings of the Conference on Nearrings and Nearfields, Hamburg, Germany. July 30–August 6, 1995). 1997, 458 pp. ISBN 0-7923-4799-4

Other *Mathematics and Its Applications* titles of interest:

L.A. Lambe and D.E. Radford: *Introduction to the Quantum Yang–Baxter Equation and Quantum Groups: An Algebraic Approach*. 1997, 314 pp. ISBN 0-7923-4721-8

H. Inassaridze: *Non-Abelian Homological Algebra and Its Applications*. 1997, 271 pp.
ISBN 0-7923-4718-8

B.P. Komrakov, I.S. Krasil'shchik, G.L. Litvinov and A.B. Sossinsky (eds.): *Lie Groups and Lie Algebras. Their Representations, Generalisations and Applications*. 1998, 358 pp.
ISBN 0-7923-4916-4

A.K. Prykarpatsky and I.V. Mykytiuk (eds.): *Algebraic Integrability of Nonlinear Dynamical Systems on Manifolds. Classical and Quantum Aspects*. 1998, 554 pp.
ISBN 0-7923-5090-1

A.A. Tuganbaev: *Semidistributive Modules and Rings*. 1998, 362 pp.
ISBN 0-7923-5209-2

M.V. Kondratieva, A.B. Levin, A.V. Mikhalev and E.V. Pankratiev: *Differential and Difference Dimension Polynomials*. 1999, 436 pp. ISBN 0-7923-5484-2

K. Yang: *Meromorphic Functions and Projective Curves*. 1999, 202 pp.
ISBN 0-7923-5505-9

V. Kolmanovskii and A. Myshkis: *Introduction to the Theory and Applications of Functional Differential Equations*. 1999, 664 pp. ISBN 0-7923-5504-0

K. Murasugi and B.I. Kurpita: *A Study of Braids*. 1999, 282 pp. ISBN 0-7923-5767-1

J.S. Golan: *Power Algebras over Semirings. With Applications in Mathematics and Computer Science*. 1999, 214 pp. ISBN 0-7923-5834-1